Introduction to Global Plate Tectonics I

Plate Tectonics Theory, Paleogeography, & the Ocean Basins

By

William A. Szary

By
William A. Szary

Library of Congress Cataloging In Publication Data

Szary, William A.
Introduction to Global Plate Tectonics I. Plate Tectonics Theory,
Paleogeography, and the Ocean Basins; William A. Szary.

Includes references.

ISBN-13 978-1500709525
ISBN-10 1500709525

1. Plate tectonics; 2. Paleogeography; 3. Historical
    Geology; 4. General Geology; 5. Earth Sciences.

Printed by CreateSpace, an Amazon Company
Original Printing August 2014
Edit Revision April 2018

Earth2Energy Educational Publishing
Port Richey FL 33618

Table of Contents

Topics: Rodinia; Pannotia; Gondwana; Proto-Tethys Ocean; Rheic Ocean; Pangea-Gondwana; Pangea; Post Pangea; Pre Modern World; Modern World; Future World.

Key Term	Definition
Nuna	A supercontinent which existed 2 billion years ago.
Crustal deformation	Folding and faulting of continental and oceanic crust as it is uplifted to form mountain belts.
Plate tectonics	A theory used to describe the shifting of the ocean basins and continents around the globe.
Rhyolites	A silica rich rock derived from volcanic eruptions originating from continental crust sources.
Andesites	A rock composed of both oceanic and continental crust sources, mixed together when extruded from volcanic and fissure type structures.
Subduction	A linear type trench or zone where oceanic crust is pulled beneath continental crust. The oceanic crust is partially melted as it is recycled back into the upper mantle
Island arcs	Volcanic islands which form a chain which trends in a linear or arcuate shape above an oceanic subduction zone, or along a continental margin where a subduction zone is positioned.
Transform boundary	A special type of fault which shifts oceanic or continental crust in a horizontal motion. Transform faults typically form at mid oceanic spreading ridges, allowing the slices of crust to move away from the ridge as new crust erupts along the fissure zone.
Suture zone	A mountain belt marking the position where one continent attached, or accreted onto another continental margin.

Introduction

Books I, II, and III are presented as a picture guide series intended to introduce the reader to a basic understanding of global plate tectonics, and the processes by which the continents were formed. In Book I, the first chapter introduces basic plate tectonics theory. The reader is presented with basic theoretical principles used to explain how continents were repositioned throughout geologic time. Chapter II (Paleogeography) presents the arrangements of the ancient continents as they drifted, collided, and rifted apart into the modern day arrangements we are familiar with today, laying the ground work for Chapter III (The Ocean Basins). Chapter III presents the geologic histories of the major ocean basins, and how they formed.

Book II addresses continental drift and the accretion of island arc and pieces of continental fragments which built out North America, Alaska, Greenland, Mexico, and South America.

Book III continues with the shaping of the continents, presenting Europe, Russia, Mongolia, China, Korea, and Indochina geologic histories.

Books II and III are descriptions of the tectonic histories of the continents. Technical terms were translated into commonly used language for easier understanding, to fit an introductory level approach to the subject matter.

The book is intended for those interested in geology and for those whom do not have a technical background on the specific terminology used to describe plate tectonics theory, or tectonic histories of continents. Some technical terms are used throughout the text.

The book is appropriate for later secondary level, and first or second year undergraduate students taking courses in global plate tectonics, historical geology, introductory plate tectonics, paleogeography, introductory earth science, and continental geology.

Some of the figures contained within the book series contain slide numbers in the upper right corner. Book figures were extracted from a video slide show series, following the same sequence presented in this book. Many of the labels, arrows, and other shapes contained in the figures were animated in the video slide show presentation. Book texts are transcribed from the slide show audio. Presentations may be viewed free of charge from Slideshare.com.

Early Precambrian Era geologic history can be traced back to the existence of a supercontinent called Nuna, some 2 billion years ago. Not much is known about the positions of the continents during the Precambrian due to repeated glaciations, erosion, crustal deformation, shifting of plates, and consumption of crustal plates by subduction processes. During the Middle Precambrian Era, some 800 million years ago, a supercontinent called Rodinia was present.

As Rodinia assembled, the conditions of the earth were known a little more precisely with bits and pieces of the earth's crust surviving the same geologic processes which destroyed the precursor supercontinent called Nuna.

Chapter I

Plate Tectonics Theory

A. *The Earth's Interior*

Plate tectonics theory is used to describe the shifting of continents around the globe. Current *plate tectonics theory* is based on the principle of a layered structure of the earth's interior, determined by energy waves passing through the earth's layers during earthquake movements around the globe. Some waves pass through both solids and liquids. These waves are called primary, or P waves, also called compression waves. Some waves only pass through solids but not through liquids. These waves are called secondary or S waves, called shear waves. Some waves only pass along the surface of the earth or along interfacing boundaries between layers. These waves are called Rayleigh, or R waves (compression), and Love waves, or L waves (shearing). Each wave type reaches seismographs at different times. The delay in timing is used to estimate layer thickness and liquid or solid properties. This is how geologists know the earth is layered.

The importance of the knowledge of the behavior of the layered earth helps to explain why earthquakes occur, why the earth's crust moved around throughout geologic history, and why it is currently moving around today.

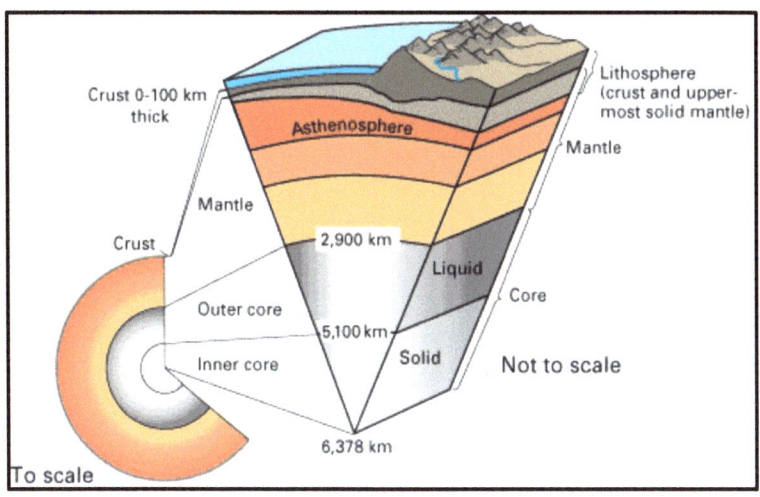

Figure 1. The earth's Interior is a layered structure consisting of a solid inner and liquid outer core; a lower and upper mantle; and asthenosphere which is part of the upper mantle. The asthenosphere circulates at a much faster rate. Lava, originating from the asthenosphere migrates upwards along fractures through the crust, and may penetrate through the crust to the earth's surface, erupting into flood basalts, cinder cones, and volcanic structures at the point where magma reaches the surface. Source: USGS Earth Interior Model.

The interior of the earth consists of a solid **inner core** composed of iron-nickel occurring at a temperature of greater than 4300 degrees Celsius (C) (**Figure 1**). The thickness of the inner core is about 1,678 km or a little over 100 miles thick.

Above the inner core lies an outer core consisting of liquid iron and sulfur at a temperature of 3700 to 4300 degrees C. The thickness of the outer core is between 2900 and 5100 km or roughly 1300 miles thick.

Above the earth's core lies the **lower mantle**. The lower mantle rocks are hot, but not molten. Rocks in the lower mantle behave as a plastic. In other words, when the rocks bend they do not return back to their original shape. When rocks return back to their original shape, they are called elastic.

Most rocks will rupture exhibiting a combination of elasticity and plasticity. The temperature near the core boundary is approximately 1000 degrees C while at the upper mantle boundary it is much cooler, at a temperature of 200 degrees C. The lower mantle's heat circulates at a much slower rate and over a wider area than upper mantle heat flow. The lower mantle ranges in thickness between 30 and 2900 km, or an average of about 1700 miles thick.

The **upper mantle** consists of partially molten rock which rises through the earth's crust along fractures and fissures. Heat flow circulates at a much faster rate and over much shorter distances. The upper mantle provides the source of magma that forms volcanoes and helps build mountain belts. It is these circulation patterns which drive the motions of the earth's crustal plates.

The uppermost part of the upper mantle which contacts the earth's crust is called the **asthenosphere**. The asthenosphere consists of very rapid convection of molten magma, or liquid rock, circulating at a fast rate.

The asthenosphere is the portion of the upper mantle which erupts at the earth's surface to form volcanic chains and uplifts mountain belts. Molten magma rises into the crust to eventually reach the surface.

The earth's crust, or lithosphere, is broken up into 13 oceanic and continental plates which slide around on top of the asthenosphere. The earth's crust ranges up to 30 km, or 18 miles thick.

B. Plate Tectonics Theory.

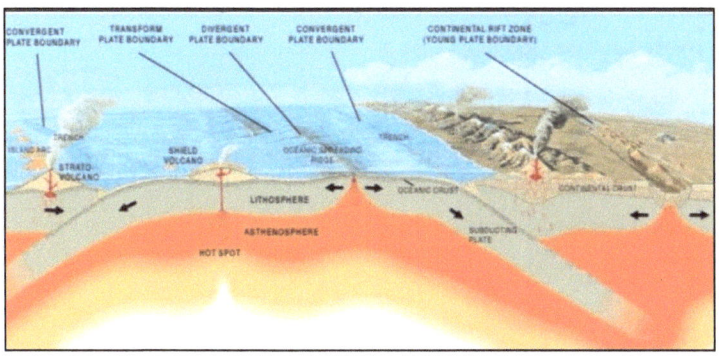

Figure 2. Plate tectonics model showing the various components of plate tectonics theory. Obtained from the USGS Plate Tectonics web site.

Definitions. Convergent plate boundaries include **subduction zones** where oceanic crust is pulled into a trench beneath a volcanic island arc or beneath the edge of a continental margin.

A convergent boundary may also occur when continental crust collides with another continent, pushing up the stationary continental edge by thrusting the mobile continental edge on top of the stationary edge (**Figure 2**).

A *divergent plate boundary* is a boundary where two separate plates are being pulled part, or rifted in opposite directions. Magma from the upper mantle erupts at the surface from fissures or from volcanic cones on the ocean floor.

Rift zones are good examples where the crust is being pulled apart. The earth's crust is being stretched due to pull apart plate motions. Continental rift systems develop within the interior parts of a continent due to stretching of the crust and pulling of the crust in opposite directions. Rifting thins the crust above the upper mantle. As the crust thins, it fractures along vertical planes of weakness allowing molten rock to rise upwards to the surface where it erupts into flowing basaltic rock. Eruptions may also build up volcanic cones which eventually produce lava, ash, cinders, boulders, etc.

The type of crust being melted in the subsurface determines the composition of lava rock that eventually is formed at the surface. For example, when an oceanic plate is pulled down beneath continental crust low in silica content, basaltic type rock is formed at the surface.

If the crustal composition consists of high silica content, the rock may form rhyolitic compositions. If there is an intermediate composition of silica present, andesites may form.

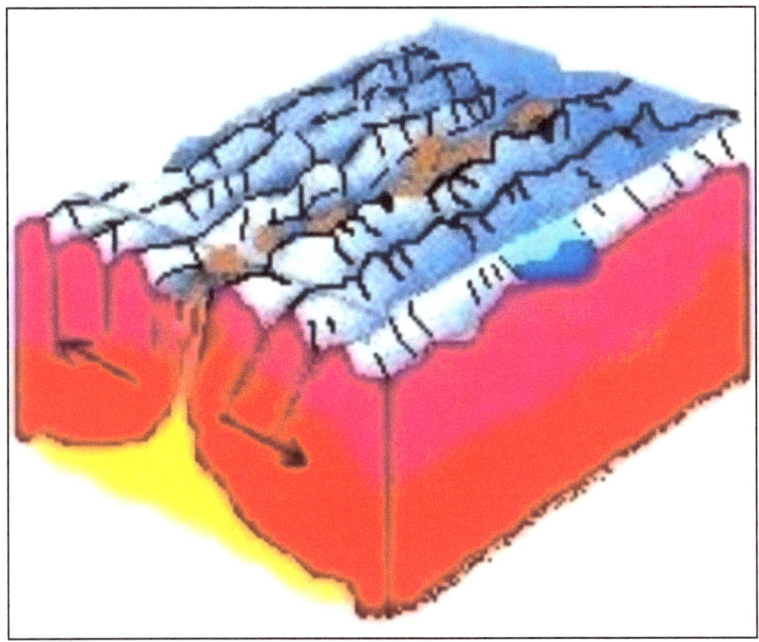

Figure 3. Spreading ridge model.

Molten magma, or lava, erupts onto the sea floor along fractures which lead from the asthenosphere to the ocean floor surface. As new lava pushes old crust away from the mid oceanic ridge, the crust breaks apart along transform faults at right angles to the mid oceanic ridge (**Figure 3**).

A *transform plate boundary* is a special boundary which forms separately from a **mid oceanic ridge,** when a mid oceanic ridge creates new crust.

The new crust is pushed away from the ridge system onto the ocean floor. The crust begins to break apart along faults, as slices of crust are pushed away.

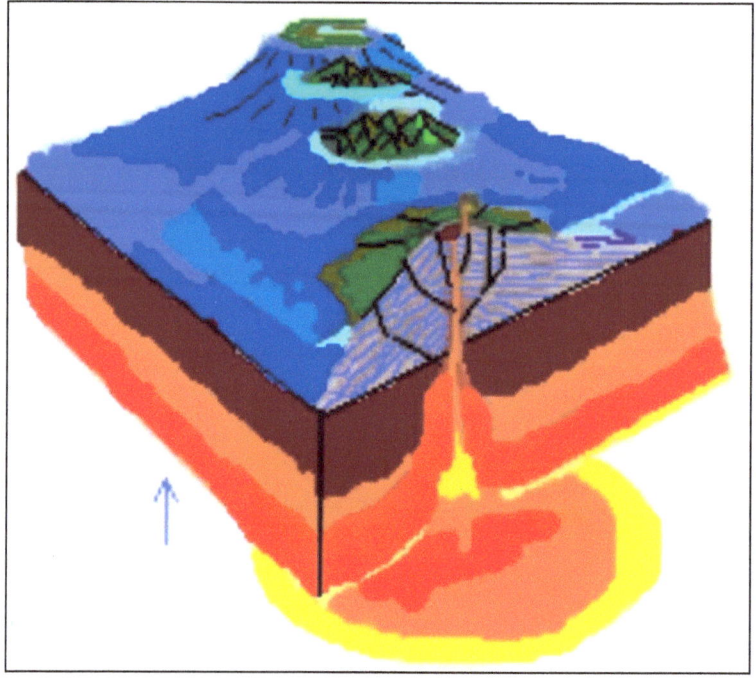

Figure 4. Hot spot model (bottom).

Transform faults form when the crust begins to break apart as the molten rock hardens and new crust near the spreading ridge pushes older crust away from the ridge breaking the crust apart into fragments. Transform faults are responsible for the process of continental drift and the widening of ocean basins.

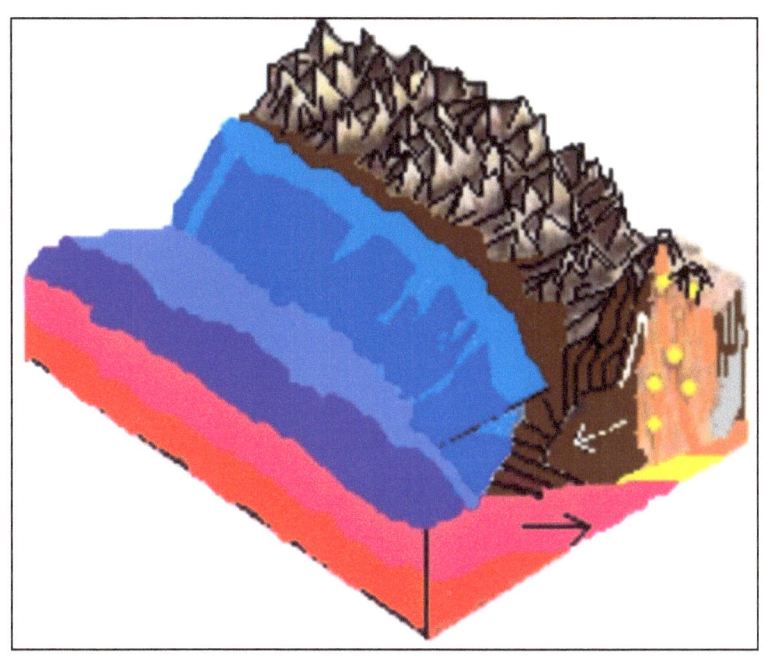

Figure 5. Subduction trench model.

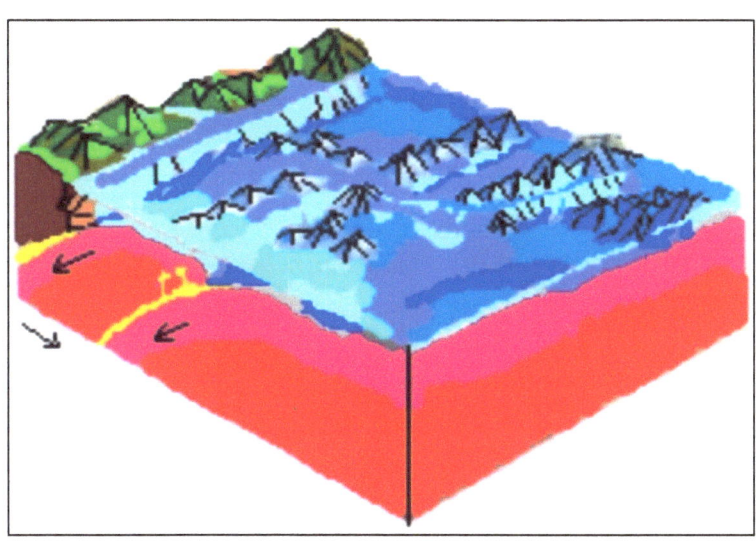

Figure 6. Subduction trench jumping (bottom).

At the **mid oceanic ridge,** new crust spills out onto the ocean floor. Molten rock, or magma, rises through the upper mantle and eventually reaches the ocean floor, erupting new crust near the ridge. Older crust is pushed away from the ridge, as the ocean basin widens.

Hot spots form over mantle plumes. Mantle plumes are areas where the mantle lies below thinner crust. Fractures in the thin crust allow magma to rise to the surface erupting continuously, while at the same time building volcanic structures on the ocean floor. Volcanic hot spots form when radioactive elements present in the earth's mantle create enough heat to melt the mantle rock. The molten rock is hotter than the surrounding mantle rock which causes the melt to rise into the earth's crust, forming an active volcano on the ocean basin floor. As the oceanic plate moves in the direction of the arrow, into the page, the hot spot begins the process of creating a new volcano on the ocean floor (**Figure 4**).

A chain of volcanoes forms marking the position of the hot spot throughout geologic time. The volcanic chain traces the path of the oceanic plate marked by older volcanoes positioned further away from the active hot spot volcano.

More recent theories suggest in addition to plate shifting, the hot spot in the upper mantle shifts as a result of moving heat circulation patterns. The Hawaiian Island Chain developed by the process of mantle hot spot shifting. The Hawaiian Chain trends in a more westerly direction than the earlier Emperor Seamount chain.

The Hawaiian Islands are approximately 400,000 years old, very young by geologic standards.

Hot spots also occur beneath continental crust. For example, the Yellowstone Hot Spot is located in Wyoming. The hot spot is responsible for producing heated ground water and steam pressures released by the famous geysers. Yellowstone is actually a collapsed volcanic structure called a **caldera**.

The Snake River Plain of southern Idaho was also formed by a hot spot mantle plume which allowed flood basalts to spill out over the landscape when the North American continent rotated.

An oceanic-continental collision results in a subduction zone trench where denser oceanic crust is pulled down under lighter continental crust. Some distance away from the spreading ridge the oceanic crust will eventually collide against a continental margin.

Because the ocean crust is cold and dense (heavier than continental rocks), the oceanic crust is forced beneath the continental crust (**Figure 5**). The oldest oceanic crust (light purple) sinks first. The younger oceanic crust (dark purple) sinks next. As the oceanic crust is pushed beneath the continental crust, it begins to melt. The oceanic crust does not melt completely. The oceanic crust melts partially.

The melted oceanic rock becomes lighter than the continental rock and begins to float upwards into the continental crust (orange).

As the melt rises, it pushes some of the overlying rock upwards and melts some of the surrounding rock changing the final rock composition. Volcanoes form where the melt reaches the surface. As the ocean crust sinks beneath the continent, the continental crust begins to break apart as it is pushed upwards into mountain ranges (**Figure 5**).

As oceanic plates subduct beneath the continent, volcanic structures on the sea floor are pulled into the subduction trench (**Figure 6**). The trench becomes clogged as the oceanic plate is jammed against the continental margin. Subduction becomes extinct and a new trench forms seaward of the former trench.

When island arcs are pulled downward into the trench beneath the continent, the arcs are equal in density to the continental crust. The arc structures and oceanic sediment plugs up the trench, forcing the oceanic plate to jump seaward. As the oceanic plate jumps seaward, a new subduction zone is created further offshore.

When small island continents (also called volcanic island arcs, for example) collide with a larger continental margin along a subduction zone, the island arc rafts into and becomes part of the new continental edge, resulting in the building out of the continental margin. The added crust is called *accretion*. The new wedged island arc causes the subduction zone to jump to the backward side of the former island arc. The basin between the new island arc and old island arc is called a *back arc basin*.

Magma begins the process of partial melting below the new subduction trench, creating new volcanic structures above the trench zone. The zone in front of the new island arc is called a *fore arc basin*.

The older oceanic crust in the back arc basin, positioned above the new subduction zone begins to push upwards, forming a new continental margin on the oceanic side of the older continental margin. The result begins the process of building out the continental margin increasing the continental land mass area (**Figure 6**).

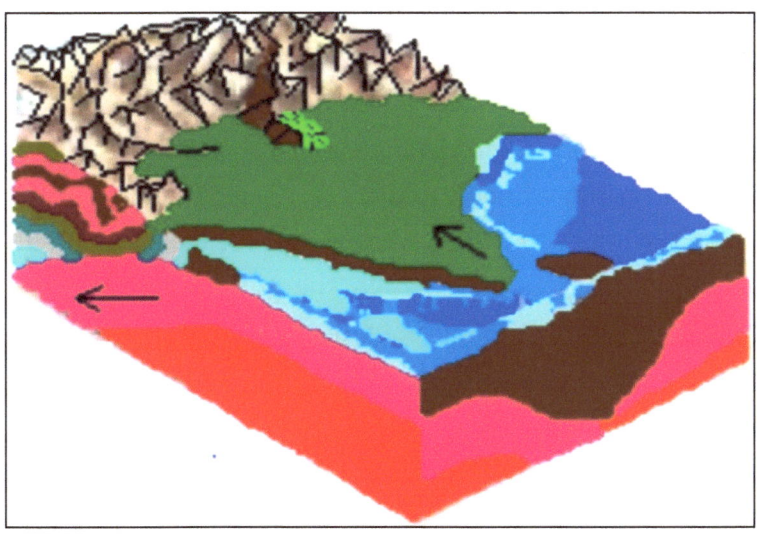

Figure 7. Continent-continent collision and obduction.

Continent-continent collisions push up high mountain belts along the suture zone, or zone of accretion where rocks of equal density meet each other. A suture zone is where two continents collide against each other resulting in the uplift of mountain belts (**Figure 7**).

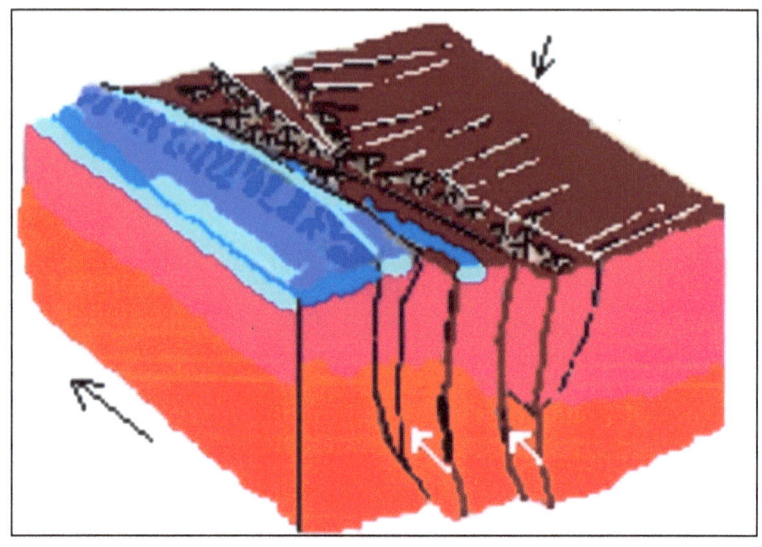

Figure 8. Transform fault boundary-continental margin collision.

When smaller island continents or subcontinents are carried on top of oceanic plates, the subducting oceanic plate continues to dive down beneath the main continent, as it normally would. When the subcontinental rocks collide, the rocks that make up the subcontinent are equal in weight and mass (or density) to the continental rocks contacting the subcontinent. The subcontinental crust is pushed skyward instead. Thrust faults form along the base of the mountain front while buried rocks are folded and faulted in the subsurface and at the surface.

Some of the buried folds push up surface, or near surface rocks into ridges. As the collision begins, the colliding continent is equal in density to the stationary continent. This causes the colliding continent to begin to rotate as it contacts the stationary continent.

The rotation pushes up mountains into a curving belt, instead of forming a linear mountain belt. This is how the Himalayan Mountains formed (**Figure 7**).

When mid oceanic ridges collide with continental margins, the ridges get caught beneath the continental crust and begin to break apart the crust into a fault system which moves crustal fragments horizontally, or laterally, in the direction of motion and at different rates.

As oceanic plates move across an ocean basin the mid oceanic ridge system may enter a subduction zone. The central rift zone becomes wedged beneath the edge of the continental margin (**Figure 8**). The rift zone faults also wedge beneath the edge, and break up the overlying continental rock so that the fractures may extend upwards to the surface. These types of faults are called San Andreas, or Transform Type Faults, named for the types found in California. They are also called lateral or strike slip faults where the crust is broken into slab sections, each moving at different rates during different times. The white arrows mark the different blocks sliced by lateral moving fault blocks and the relative direction of motion. When small fault segments intersect with each other, a fault zone is formed. Lateral type faults usually form a continental rift zone where the landscape forms a linear depression (**Figure 8**).

Finally, the portion of the continent that continues to remain stable and is either unimpacted, or slightly deformed by subduction, collison, folding, and faulting is called a **craton**. The craton usually consists of Precambrian Era rocks, typically igneous or metamorphic rocks that make up the stable part of the continent. A craton is also called a **continental shield**. These terms are used interchangably.

Chapter II

Paleogeography

Paleogeography is the study of the ancient geographic positions of the earth's continents and ocean basins. These studies are based on fitting continents together like a puzzle, using shapes of the present day continental margins. Matching the ages of the rocks positioned along continental margins and within plate interiors are also used to reconstruct continental margins. The Deep Sea Drilling Project provided rock samples used to reconstruct ocean basin histories.

In this chapter, paleogeographic reconstructions are provided through a snapshot of globes which show the former positions of continents beginning with the Precambrian Era, about 800 million years ago, leading up to the modern world as we know it. Reconstructions are vague prior to the Precambrian Era due to rocks being destroyed by erosion, glaciation, burial beneath younger rocks, and destruction by crustal plate consumption along subduction zone trenches. Limited history for some continents remains where older rocks are still exposed at the surface. Ancient geologic history is presented in greater detail in Chapter III, detailing the ocean basins.

Two future models are presented in this chapter. The future models show the positions of the continents after 50 million years and 250 million years into the future.

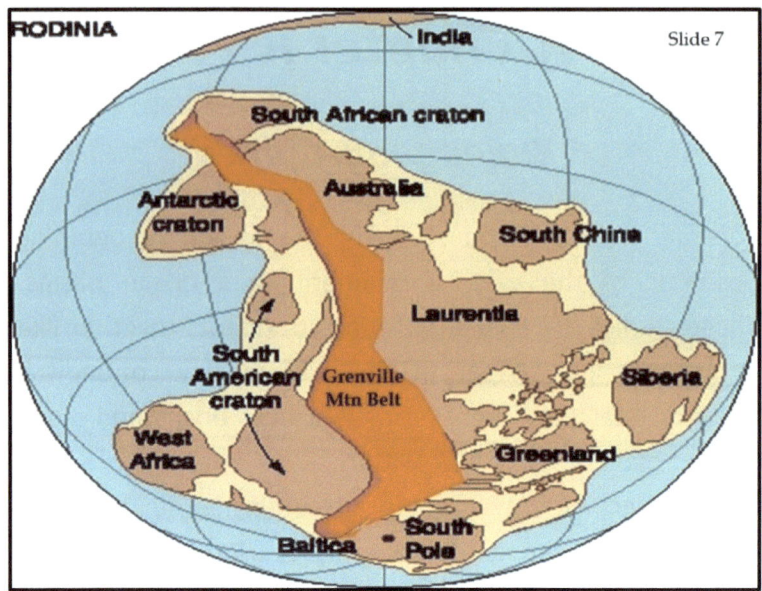

Figure 1. Rodinia formed after the breakup of Nuna. The dark red belt represents the Grenville Mountains which uplifted when Laurentia collided with Australia, Baltica, and South America.

Rodinia occupied the earth during the Late Precambrian Era, 800 million years ago. The center red belt represents the Grenville Mountain Belt, formed as the supercontinent was assembled some 1.3 to 1.0 billion years ago (**Figure 1**). Between 760 and 726 million years ago, plates were relatively small. Upper mantle heat flow was rapidly convecting, causing plates to slide around very fast.

Pannotia formed after Rodinia broke apart, 650 million years ago (**Figure 2**).

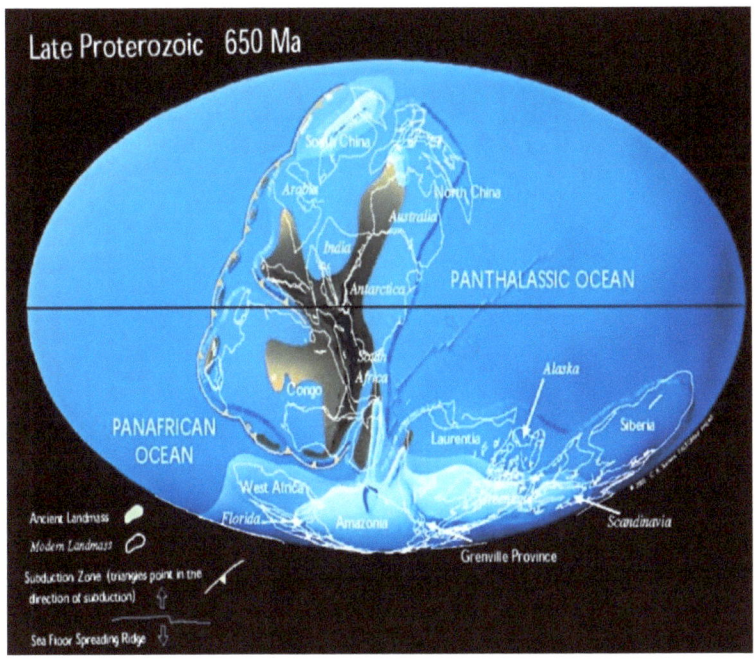

Figure 2. Pannotia formed after Rodinia broke apart. Laurentia, Amazonia, Baltica, Siberia, and Africa were covered with ice at the South Pole. Antarctica, India, Australia, and the Congo made up the supercontinent which extended from the southern latitudes through the Equator, into the northern latitudes.

The sub-continental plates rearranged themselves into a more linear shape with branches extending outward towards the northwest along the western continental margin. Precambrian rocks observed in Australia, India, Malaysia, Antarctica, South Africa, and the Congo were used to reconstruct the supercontinent. Glaciers covered the northern tip. Portions of South China near the North Pole, and portions of West Africa, Amazonia, Greenland, and Scandinavia were covered by glaciers at the South Pole.

25

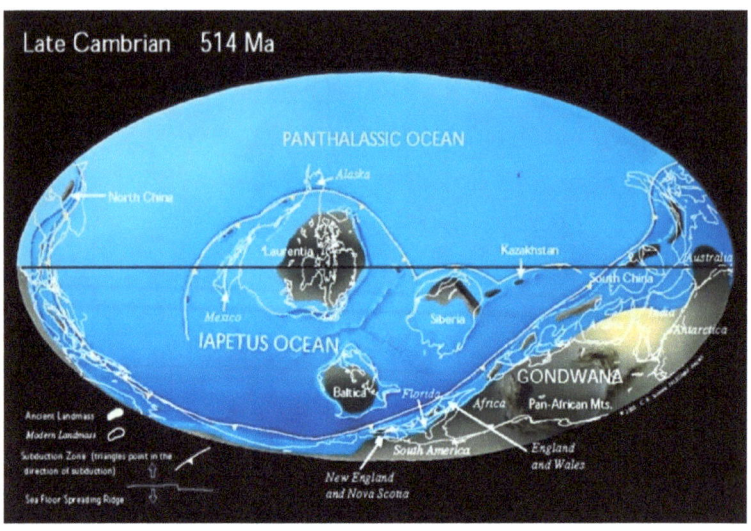

Figure 3. Gondwana formed as Laurentia, Baltica, and Siberia broke away and drifted north into the Equatorial region. The ice sheets melted at the South Pole. Gondwana remained emergent.

Gondwana and Laurentia were separated following the breakup of Pannotia, 514 million years ago (**Figure 3**). During the Early Paleozoic Era, into the Cambrian Period (512 million years ago), Laurentia, Siberia, and Baltica broke away from the Gondwana mainland. The Iaepetus Ocean opened up southwest of Laurentia. Spreading rift zones were forming to the southeast of Laurentia and an expansive subduction zone was present along the northern Gondwana coast, stretching across the entire globe. A smaller subduction zone was also located to the north of Laurentia, stretching east to the Gondwanan subduction zone.

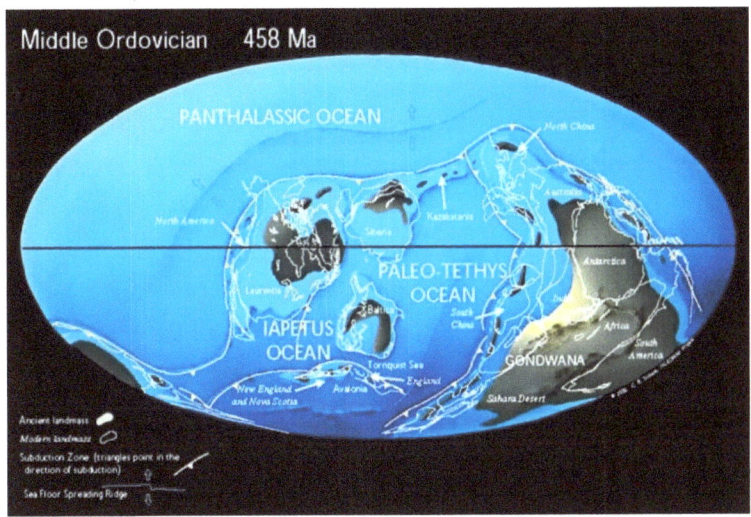

Figure 4. The paleo-Tethys Ocean opened up between Laurentia and Gondwana. Smaller oceans developed between the continental plates. The continents were rearranged around the globe.

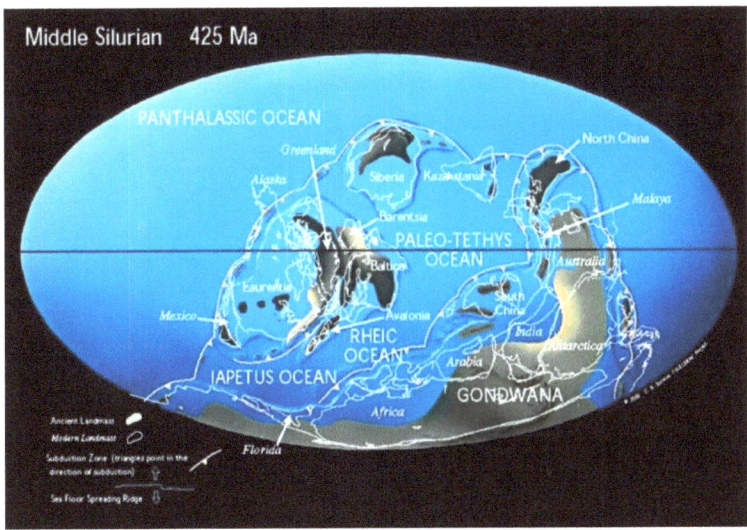

Figure 5. As island arc systems developed, and the larger plates collided, ocean basins became more numerous. Ice sheets were absent at the South Pole.

The Tethys Ocean separated Gondwana from Laurentia, 458 million years ago forming the *proto-Tethys Ocean (Figure 4)*. The earth's crust broke up into distinctive plates. Northern Laurentia (Canada) was positioned near the equator.

Gondwana migrated northward and Antarctica was positioned at the equator. The proto-Tethys Ocean opened up between Laurentia and Gondwana. Siberia migrated towards the North Pole. Baltica was positioned in the center southern hemisphere. New England was located near the South Pole. A distinct continental shelf developed around the major continental margins.

A spreading ridge system was positioned in the northern hemisphere beneath the Panthalassic Ocean. A corresponding subduction zone surrounded the major continents. Volcanic island arc systems were located above the subduction zones (**Figure 4**).

The **Rheic Ocean** separated Laurentia-Baltica from West Africa, 425 million years ago (**Figure 5**). Greenland accreted onto northeastern Laurentia (Canada). Baltica collided with eastern Greenland. Siberia was located near the Arctic Circle.

Small island continents occupied a continental arc throughout the central and southwestern US. The first island continent appeared in southwest Mexico.

North China broke away and drifted north of the equator from northeast Gondwana. Australia was located near the equator. The Panthalassic Ocean occupied the northern portions of the globe. The proto-Tethys Ocean remained along the equator separating the northeastern parts of Gondwana from Baltica, Greenland, and Laurentia. The Iaepetus Ocean was located south of Laurentia and the Rheic Ocean occupied the eastern margin of Laurentia. Many of the plates were smaller in size surrounding each of the major continents (**Figure 5**).

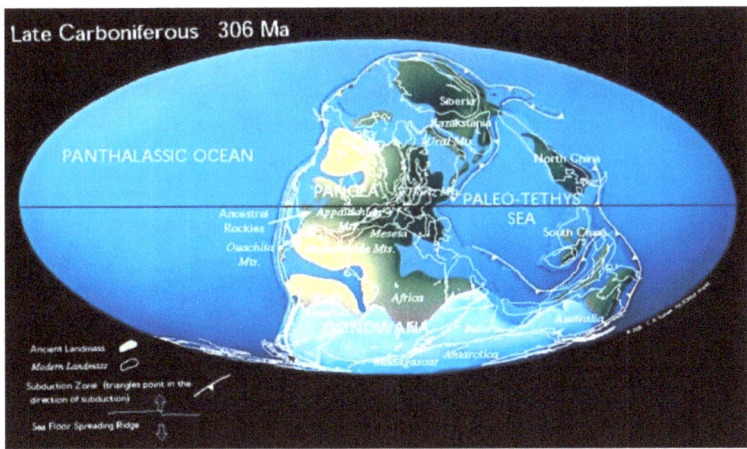

Figure 6. Pangea collided with Gondwana. Many of the smaller ocean basins closed leaving behind larger oceans surrounding the supercontinents. Ice sheets reappeared covering a much larger land area in the southern latitudes. Siberia started to collide with other smaller island arcs.

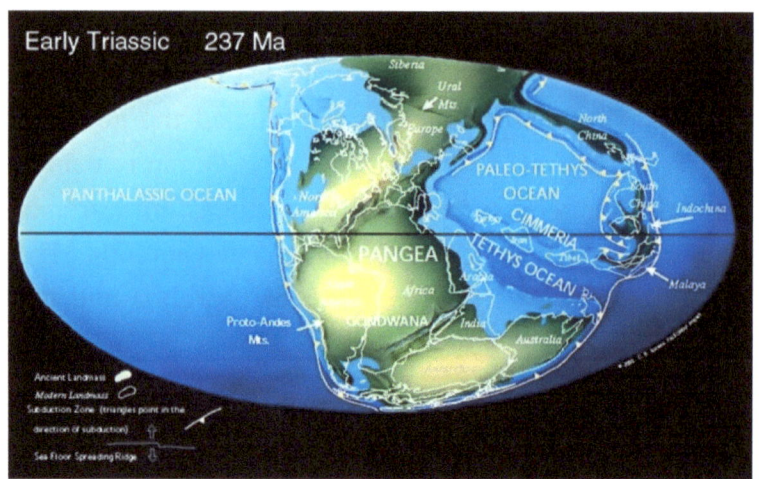

Figure 7. Pangea started expanding in area. Siberia collided against Europe. The Pacific Rim began to form as the Korean Peninsula and the Japanese Islands formed along the western Pacific margin. Glacier ice retreated in the southern latitudes.

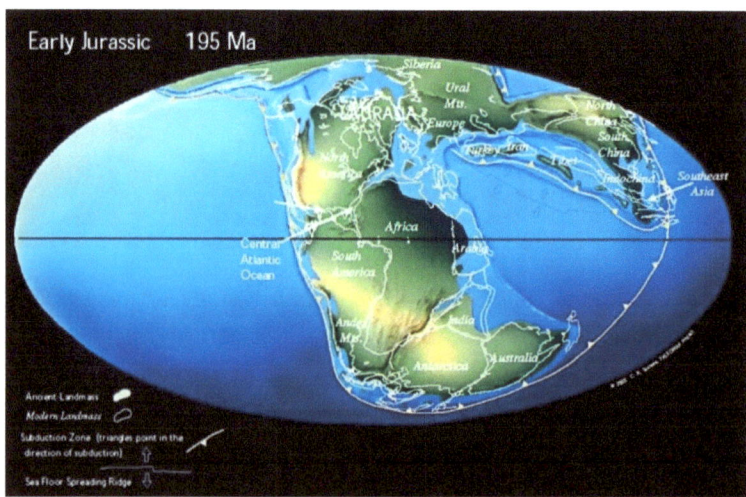

Figure 8. Eurasia began to break away from Laurentia. Pangea started to split apart. Gondwana began to separate from Laurentia. Asia began to take shape. Ice sheets were absent at the poles.

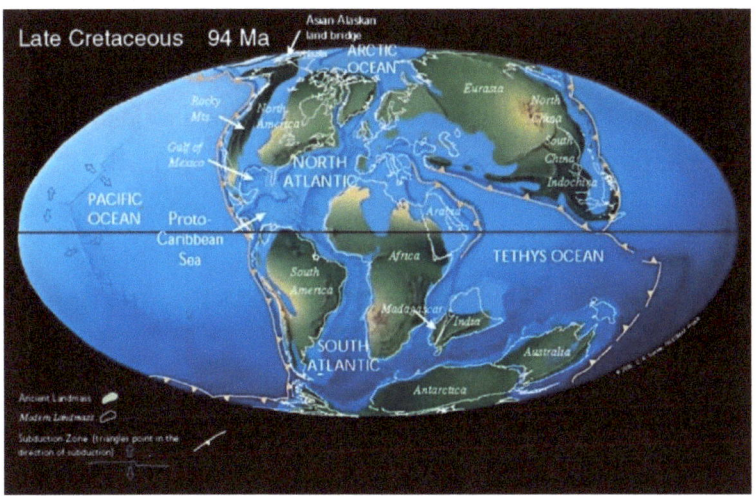

Figure 9. The pre-modern world began taking shape. Sea levels rose covering the western interior parts of North America, interior Europe, and northern Africa.

Pangea and **Gondwana** collided to form Pangea, 306 million years ago (**Figure 6**). Most of the northern portions of Gondwana collided with Laurentia to form Pangea. The Appalachian, Ouachita, and ancestral Rocky Mountains formed from these collisions during the Pangean supercontinent assembly.

North China drifted north to the eastern central northern hemisphere. South China was an island located near the present day Australian continent. The Panthalassic Ocean occupied the present day Pacific Ocean region. The Tethys Sea occupied most of the eastern region. The Iaepetus Ocean closed when South America drifted towards the West African coastline. Much of the Gondwana continent was covered by glacial ice near the South Pole (**Figure 6**).

During the Early Triassic Period, Europe and Siberia collided together forming the Ural Mountain suture zone in central Russia (**Figure 7**). The Korean Peninsula and Kamchatka was attached to eastern Siberia, forming the Asian mainland. Eur-Russia collided with Laurentia forming the Laurussian supercontinent. Gondwana also collided with Laurentia forming the pre-Pangean supercontinent.

In the eastern Paleo-Tethys Ocean, an island arc system formed above a regional subduction zone which extended along the western Panthalassic Ocean. The Paleo-Tethys Ocean was separated from the Tethys Ocean by an island arc system called Cimmeria. Cimmeria consisted of portions of Turkey and Iran.

Geologist use the prefix Paleo- to mean the Paleozoic aged Tethys Sea, since the sea persisted throughout geologic history. Repeated opening and closing of the Tethys Ocean continued into the Mesozoic and Cenozoic Eras.

Malaysia and Indochina made up the larger island arcs along this trench system which extended from eastern Kamchatka to the South Pole. The trench system continued around the Australian-Antarctica continents, north along the South America portion of Gondwana, and off the North American west coast (**Figure 7**).

Pangea was assembled, 195 million years ago (**Figure 8**). The North and South Poles were free of glacial ice, raising global sea levels.

The central Atlantic Ocean began to open along a rift which developed between the eastern North American and northwestern African margins. South America was attached to the western African continent. Arabia was attached to the East African coast. India was attached to Antarctica, along with Australia.

Greenland was wedged between Europe and the North American continents. All three continents were accreted together as a single continent. Siberia was sutured to Eastern Europe by the Ural Mountains. China was assembled along with Southeast Asia attached to eastern Siberia. A subduction zone encircled Pangea and Laurasia. A mid oceanic ridge system was positioned in the Tethys Sea accompanied by a subduction trench and a volcanic arc system, positioned off shore from the Eurasian supercontinent (**Figure 8**).

Pangea began to break apart (Figure 9). About 94 million years ago, the North American, South American, and African continents separated as the Atlantic Ocean continued to spread apart along a mid-oceanic rift system.

Greenland was beginning to separate from Northern Canada by rising sea levels. Glaciers disappeared from the polar region during the Middle Jurassic Period.

Eurasia was moving eastward while the Ural Mountain suture zone continued rising. China and Southeast Asia accreted together to form Eurasia. The entire continent rotated clockwise towards the equator.

India broke free from southeast Africa and began drifting to the northeast with Madagascar attached to the Indian subcontinent. Australia remained attached to Antarctica.

The Gulf of Mexico began opening. The Western Interior Seaway connected the Gulf of Mexico to the Arctic Ocean along the interior west of North America. A subduction zone formed around the Pacific Ocean rim along the eastern Asian continental margin off the east coast of the African continent and off the east coast of Antarctica-Australia. The Pacific Rise developed as a spreading ridge center in the central Pacific Ocean.

The Atlantic Ocean was also widening. South America began drifting away from Africa. Africa was beginning to rift along the central western and central northern regions. Seas occupied the Saharan Desert region (**Figure 9**).

Following the breakup of Pangea the continents spread apart along mid oceanic ridge systems (**Figure 10**). A subduction zone formed around the north and eastern rim of the Caribbean Basin. Madagascar broke off from the northwestern Indian sub-continent. India collided with South China uplifting the Himalayan Mountains. Mountain uplift resulted in the Tibetan Plateau region uplift just inland of the mountain chain. Southeast Asia, Indonesia, and Malaysia formed above a subduction zone lying off the northern Australian coast. The Hawaiian Island chain continued to form over a hot spot located in the central Pacific Ocean.

The Gulf of Aden and Red Sea Rifts formed, as Saudi Arabia began to pull away from northeast Africa. Glaciers occupied the Arctic Circle, Greenland, and Antarctica continents.

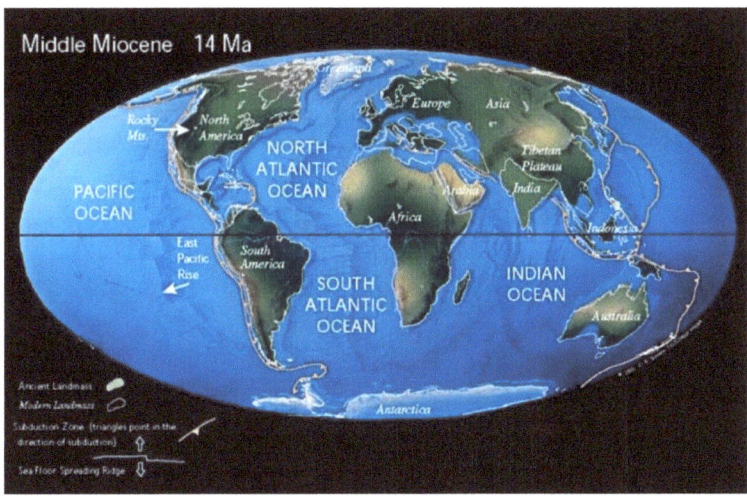

Figure 10. The pre modern world began to take shape. The Atlantic Ocean, Indian Ocean, and Pacific Ocean were well developed. Antarctica was covered in ice along with Greenland and the Arctic Circle.

The **Modern World** is currently broken up into 13 major and minor tectonic plates marked by mid oceanic ridges and subduction zones (**Figure 11**). Rift zones are forming where new plates are separating along edges of continents and within continental interiors. Volcanic chains are forming over hotspots on the sea floor, within continental interiors, and above subduction zones.

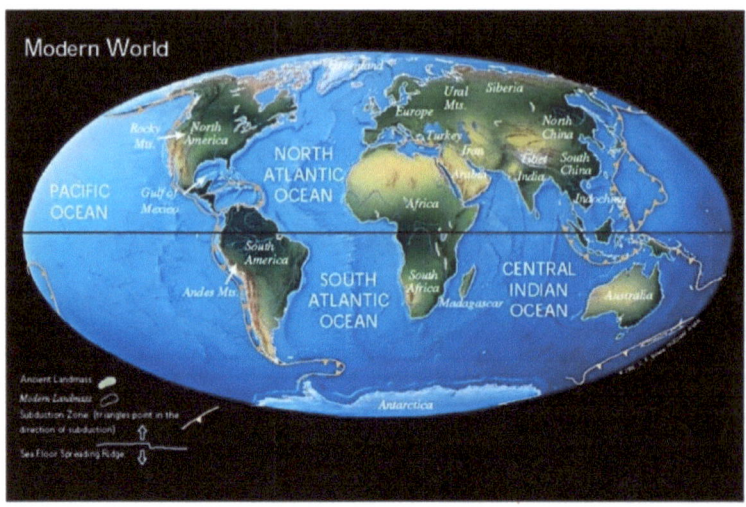

Figure 11. The modern world was formed as we know it today consisting of 13 major plates which continue to shift around the globe.

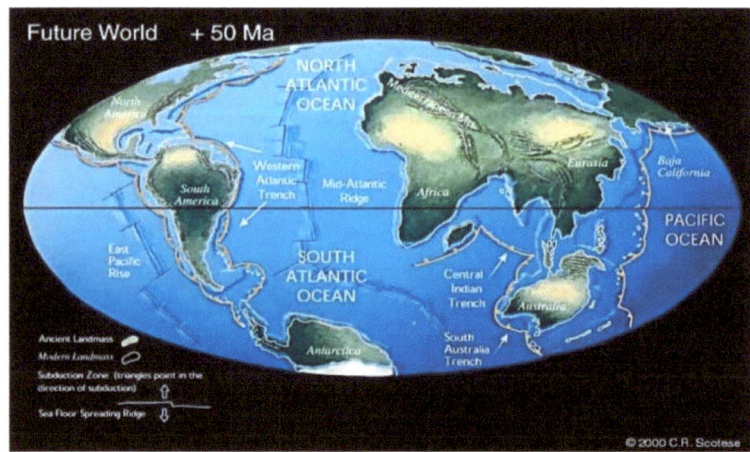

Figure 12. The future world as it would appear 50 million years from now places Africa against the European and southwest Asian margins. North Russia and China break away from northern Europe, and New Zealand collides with northeast Australia. Glaciers partially cover Antarctica.

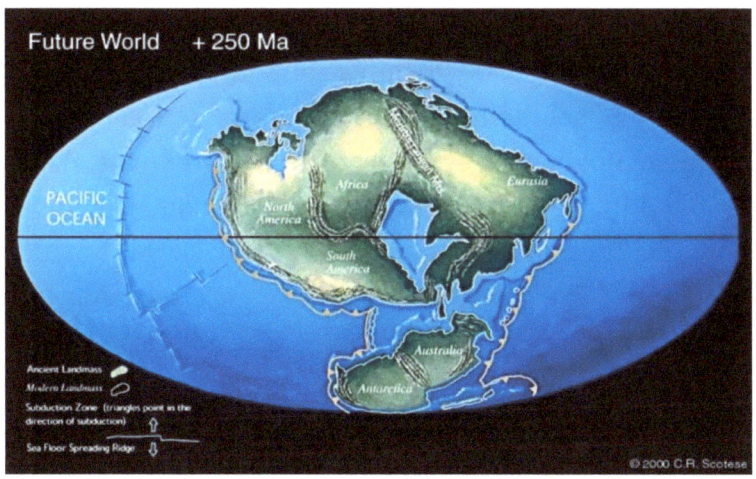

Figure 13. New Pangea forms 250 million years into the future. North America collides with South America accreting onto western Africa. Antarctica and Australia collide again at the South Pole. Glaciers are absent at both poles.

The present world continues to shift as oceanic plates continue to spread at mid oceanic ridge centers beneath the oceans, and within continental interiors. Oceanic plates continue to subduct beneath continental margins. Volcanic chains and mountain belts are created along continental margins while eruptions also occur within the interiors of oceanic plates and continental interiors. Volcanism also occurs where the crust is being pulled apart to provide pathways for molten rock within the asthenosphere to rise up and erupt at the surface. Continent-continent collisions are also pushing up mountain ranges where suture zones are being deformed by folding and faulting (**Figure 11**).

About 50 million years into the future, Africa, Europe, and Southeast Asia collide together to form a suture zone along the southern European margin (**Figure 12**). Northern Russia and China break away from the Baltic region, drifting northeast as a new subduction zone forms along the southern Arctic margin. Greenland drifts across the Arctic Circle towards the new Russian subcontinent. Australia collides with New Zealand along the northeastern Australian margin. Indochina also becomes part of the accreted land mass. Antarctica drifts west towards the South American peninsula. The northern half of the Antarctic continental glacier retreated due to a change in latitudinal positioning. South America drifts north, compressing the Caribbean basin against the North American plate.

Moving forward 250 million years from now, the New Pangea assembles as South America collides with North America by continental drifting to the east. The Atlantic Ocean closes (**Figure 13**). The Americas collide with the western African coast creating another supercontinent. Australia and Antarctica collide together in the southern hemisphere to form a separate supercontinent positioned north of the South Pole. A subduction zone system rims both continental land masses. A continuous spreading ridge center forms in the mid Pacific Ocean.

Chapter III

Global Tectonics Evolution:
The Ocean Basins

In this chapter the geologic history of the ocean basins is presented to provide the reader with a basic understanding of how the continents were separated throughout geologic time. Each ocean basin is discussed individually. The reader should keep in mind the earth is a dynamic place and that all motions are continuous and occur simultaneously together. This section is divided into nine subsections, presenting each ocean basin as a separate discussion although all continents were in motion for all ocean basins at the same time during development.

Pangea was assembled prior to the formation of the Atlantic Ocean Basin (**Figure 1**). The Appalachian Mountains occupied the eastern margin of Laurentia, including portions of Western Africa.

The North American, African, and South American plates were sutured (attached by collision where mountains belts were uplifted) along the Appalachian Mountain belt during the Late Devonian (360 million years ago) through the Late Pennsylvanian Periods (300 million years ago). South America was already separated along a narrow seaway between the South American section of southwestern Laurentia and northern Gondwana.

Rifting began 230 million years ago. The Appalachian mountain belt suture between Laurentia and Gondwana began to collapse.

A. The Atlantic Ocean Basin

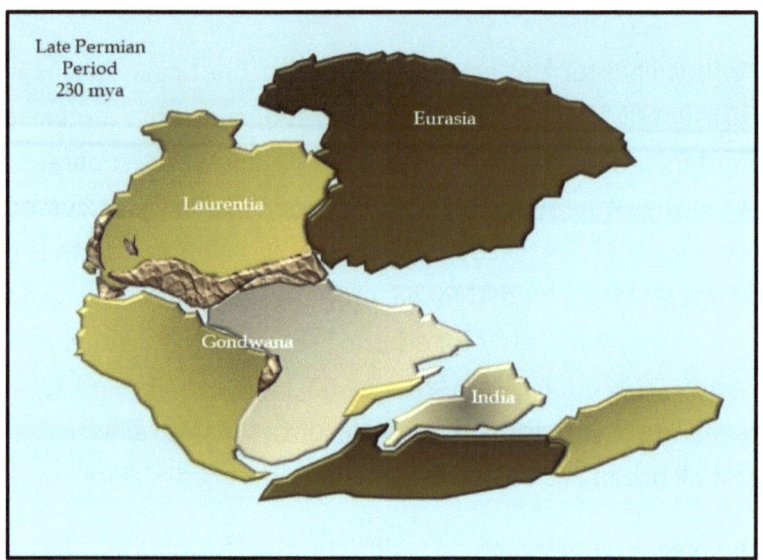

Figure 1. The Assembly of Pangea.

Laurentia began to pull away from Gondwana opening up a rift between North and South America (**Figure 2**). The southern segment of Mexico rifted away from the peninsula. The remaining peninsular area was attached to southwestern Laurentia. Rifting began between northern South America and western Africa expanding eastward. South America continued to separate from Africa. Embayments began to flood toward the southern Laurentian continent, along the Ouachita Mountain front. Subsidence continued collapsing the Appalachian belt.

The eastern Appalachian suture zone began to flood. Baltica began to pull away from northeastern Laurentia.

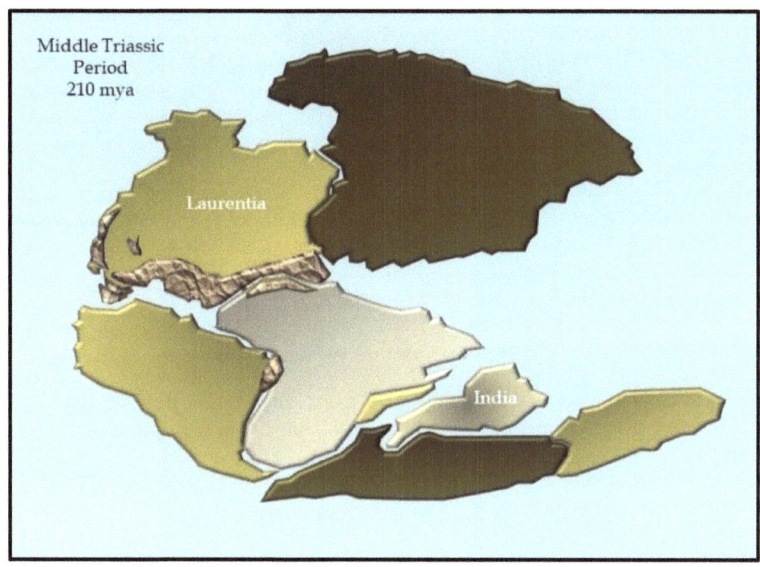

Figure 2. Rifting of Laurentia from Gondwana.

Pangea began to break apart beginning the opening of the South Atlantic Ocean between the North and South American continents (**Figure 3**).

Baltica, Eurasia, Iberia, and portions of the African northern continental margin rifted from northeastern Laurentia. Separation of the continents opened up the northern proto Atlantic Ocean. South America continued to separate from southern North America. Rifting occurring along the southwestern Laurentian margin began the development of the Mexican peninsula region.

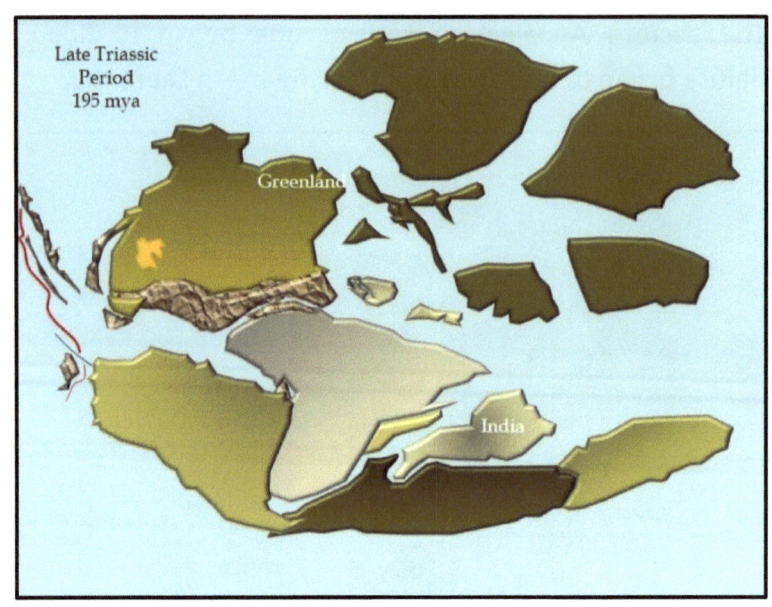

Figure 3. Opening of the South Atlantic Ocean.

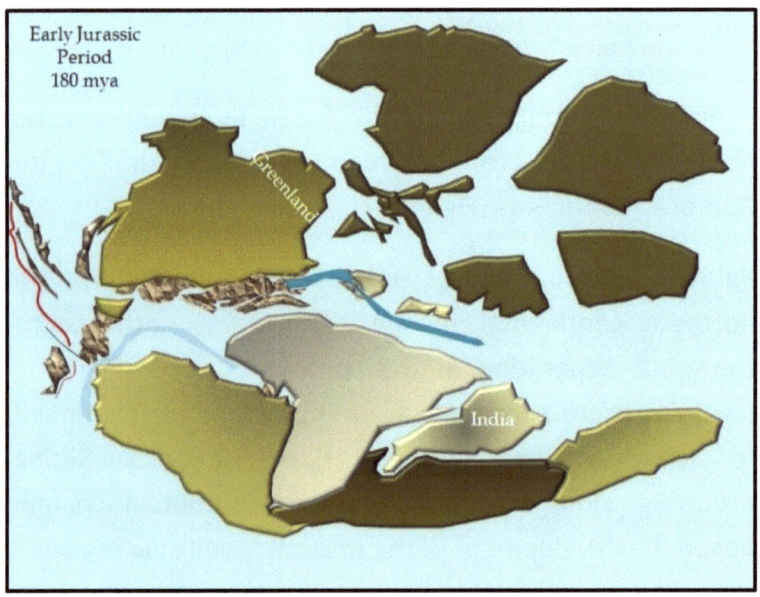

Figure 4. Rifting of the Central Atlantic Ocean.

42

Continental separation collapsed the Appalachian Mountains, promoting subsidence and erosion of the range. Africa and North America gradually widened (**Figure 3**).

As Gondwana rifted away from Laurentia, opening of the Central Atlantic Basin continued (**Figure 4**). A mid oceanic ridge began developing in the North Atlantic Ocean. Rifting deepened along the northern African margin. Africa continued separating from North America along an expanding mid oceanic spreading ridge center (dark blue strip). South America remained attached to Africa, while the South Atlantic Ocean continued widening and deepening. In the Mid Atlantic section, the spreading ridge began to expand between South America and Africa (light blue strip).

Continued subsidence and erosion of the Appalachian Mountains occurred. Rifting between Africa and North America continued to widen, opening up the Central Atlantic Ocean. The Mexican peninsula reassembled. Mountains rifted from northwest South America, colliding with the Mexican Peninsula (**Figure 4**).

The Appalachian Mountains continued eroding while the Atlantic Basin continued to widen (**Figure 5**). The mid oceanic ridge began extending into the central Atlantic Ocean. Africa and South America continued moving away from North America remaining attached as Gondwana.

The Appalachians were eroded down to a flattened, submerged plain except for a small continental fragment which rifted away from the Gulf of Mexico region along the southern North American continental margin. The northern segment of the Atlantic mid oceanic spreading ridge extended towards the central ridge system.

A major rift developed in the central western North American craton, opening up a subsiding basin which allowed the Arctic Sea to invade into the central western portion of North America. The South, Central, and Northern Atlantic Ocean basins were connected by the Middle Jurassic Period (**Figure 5**).

Figure 5. Appalachian uplift.

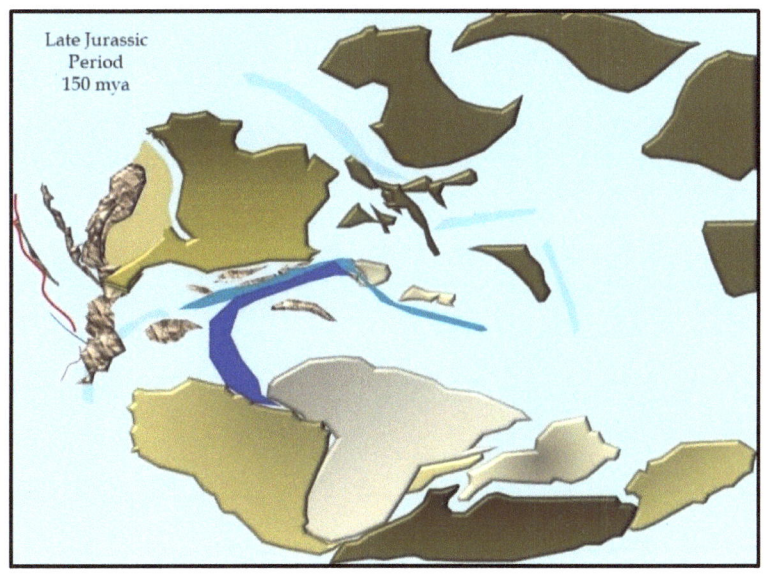

Figure 6. South African Rift.

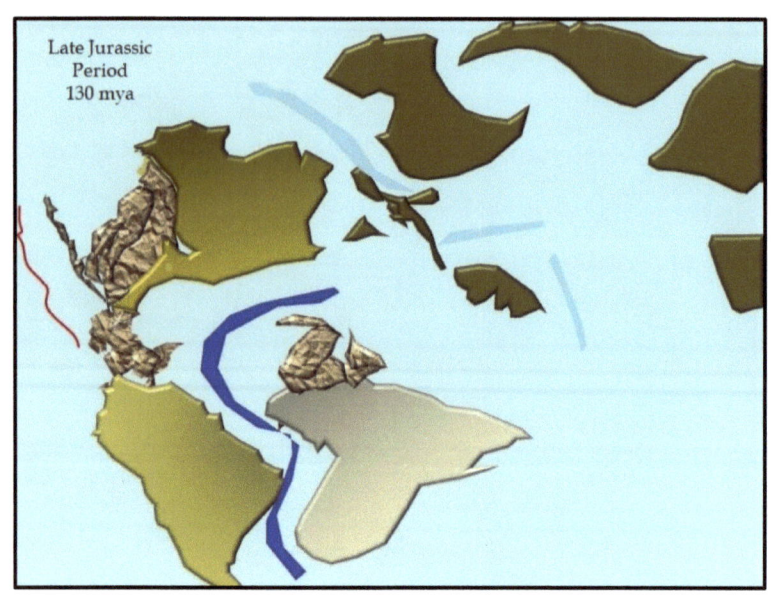

Figure 7. Gondwana Rifting.

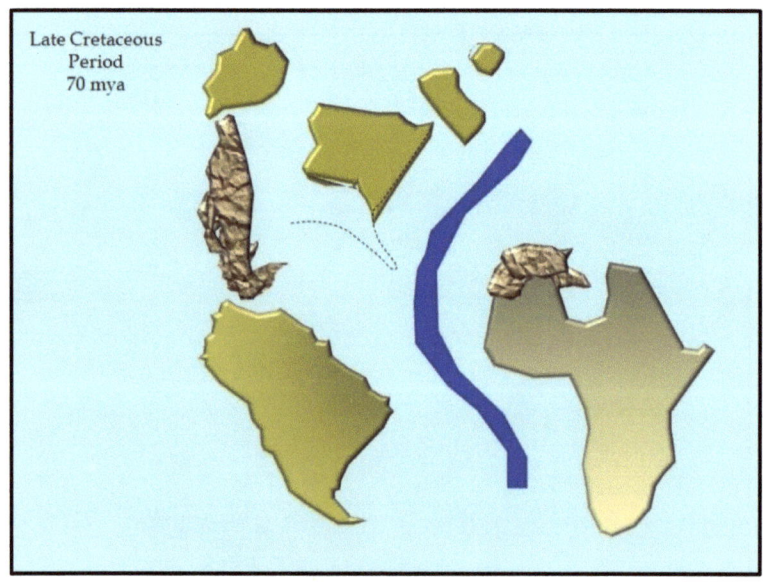

Figure 8. Laurentian Sea Level Rise.

The South Atlantic Ocean began to extend into the rift zone between South America and Africa (**Figure 6**). The mid oceanic ridge curved to the south (dark blue strip), extending into the rift occurring between South America and Africa. Eurasia rifted apart. Iberia and smaller island fragments continued drifting east. Africa and South America continued moving away from North America, remaining attached to Gondwana. The Atlantic Ocean continued to deepen.

Collisions between island arc terranes, erosion from the western North American mountain belt, narrowed the inland sea by filling in the subsiding basin with sediments. Sediments were transported eastward from the eroding mountain front. The Panamanian isthmus continued developing while island arcs collided with the southern Mexican peninsula. The Gulf of Mexico began forming and the Yucatan began drifting south towards the Mexican peninsula (**Figure 6**).

South America rifted away from Africa, opening the South Atlantic Ocean (**Figure 7**). Mexico and Middle America extended to the northern South American continental margin. The mid oceanic ridge system was well developed throughout the Atlantic Basin. North America started rotating as the Yucatan Peninsula collided against the Mexican Peninsula. South America started rotating clockwise away from Africa. As South America tugged at western Africa, the Central African rift began extending eastward.

Northern Africa rifted away from the northwestern continental margin. The Atlantic Ocean widened, establishing a spreading center. The Yucatan peninsula collided with the Mexican peninsula.

North America rotated counterclockwise towards the west (**Figure 8**). Shallow seas invaded the central and southern North American continent due to sea level rise. Africa reached its present position as a rift formed in the central northern region of the continent. The Atlantic Basin formed a mid-oceanic ridge system at the same time.

Greenland rifted from northeastern Canada. Smaller continental fragments drifted from the Canadian margin at the same time.

North America continued rotating counterclockwise. As North America rotated, a shallow inland sea invaded most of the western interior parts of the continent. The western mountain belt also rotated and uplifted, cutting off the Pacific Ocean from the inland sea. As South America and Africa drifted away from each other, the Atlantic Ocean widened and the spreading ridge continued erupting on the ocean floor (**Figure 8**).

Figure 9. The central African Rift formed as a transform fault which extended into the western African continent, expanding north into the central Saharan Desert (white line). Basalt flows erupted along the continental margins of western Africa and the eastern Amazon (brown shapes) resulting from a rift which separated the continents.

By the end of the Mesozoic Era, 65 million years ago, the Atlantic Ocean was completely formed (**Figure 9**). The central African rift extended from the southwestern continental margin curving through the northern Sahara Desert region. Basaltic dikes and faults developed along the Equatorial Atlantic margins from the separation of South America from Africa. The continents were positioned very close to where they are recognized presently.

Figure 10. The modern day Atlantic Basin shows the former outline of the former Pangean supercontinent where rifts separated the North American, South American, and African continents.

The present day Atlantic Basin mid oceanic ridge system and outer continental shelf patterns trace the former Pangean supercontinent where the continents were formerly joined together (**Figure 10**).

The patterns allow the separated continents to be reconstructed by fitting the margins together. The age and patterns of mountain belts along the margins were also used as evidence for supporting an argument for the reassembly of Pangea.

B. The Caribbean Ocean

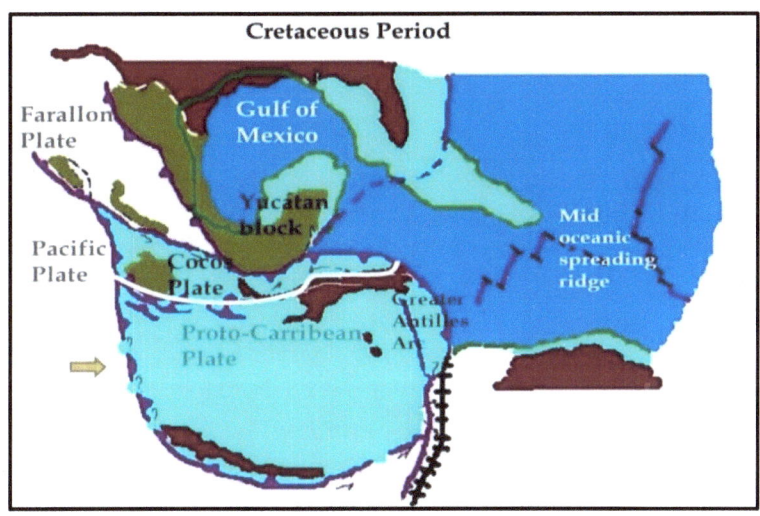

Figure 11. The Caribbean Basin forms.

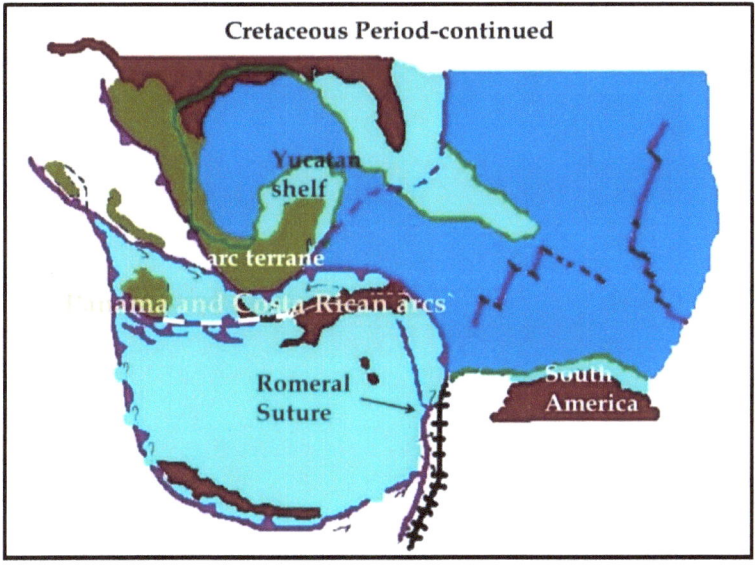

Figure 12. Island arc collisions.

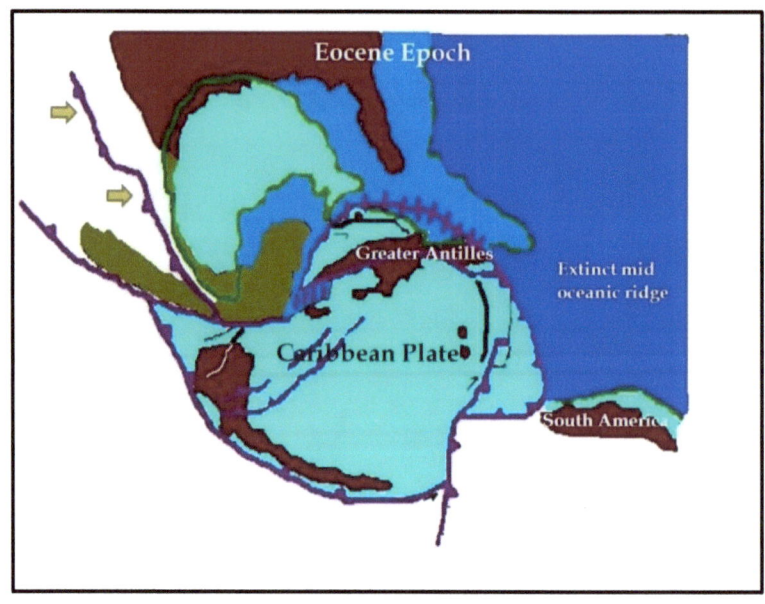

Figure 13. Greater Antilles Uplift.

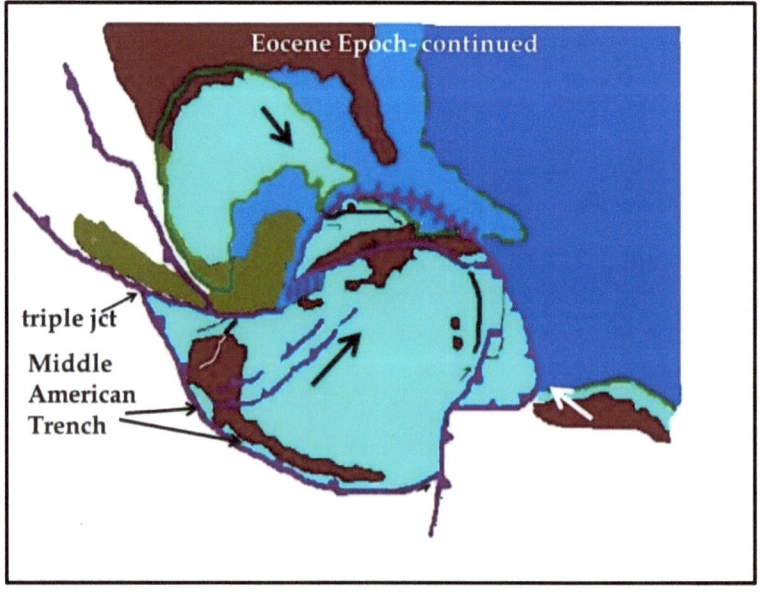

Figure 14.Caribbbean plate migration.

The Caribbean Ocean became a separate basin when a subduction zone formed between the Farallon Plate and the Caribbean Plate (**Figure 11**). During the Early Cretaceous Period, the Caribbean plate was probably part of the Pacific Plate. The Caribbean Plate separated from the Farallon Plate when the Cocos Plate formed.

The Farallon Plate moved NE relative to North America. Igneous intrusives and volcanics developed in the Greater Antilles Arc. Convergence between the Farallon Plate and the proto-Caribbean plate produced a subduction zone separating the two plates.

A single mid oceanic ridge system developed between the Yucatan and South America. South America continued separating from North America. As spreading ended in the Gulf of Mexico, the Yucatan block collided with the North American plate.

Subduction of the Cocos Plate along the Middle American Trench formed the present day boundary between the Cocos and Caribbean Plates. Between the Jurassic and Early Cretaceous Periods, Cordilleran Mexico migrated SE colliding with South America during the assembly of western Pangea (**Figure 11**). A cordillera is defined as a group of mountain ranges separated by valleys, lakes, etc.

Volcanic eruptions occurred along the margins of the Caribbean Basin (**Figure 12**). Island arcs collided with Middle America. The basin received sediments shed from the rising islands.

Deformation and uplift progressed slowly along the proto-Caribbean margin. Uplift began in the west along the southern Yucatan shelf, resulting in a collision between a north facing arc terrane and the southern Yucatan terrane. The Panamanian and Costa Rican arc produced igneous intrusion injections into oceanic basement crust. Marine sediments covered the crust marking the separation of the two plates by subduction, about 84 million years ago.

From the Late Cretaceous Period forward, the Caribbean Sea margins were of the passive Atlantic type (without volcanic activity) receiving sediments shed from island margins.

The eastern and southern Yucatan block and northern South America belonged to a part of this margin. Uplift also occurred along the South American Western Cordillera forcing the closure of the Romeral Suture (**Figure 12**).

The Caribbean Plate migrated to the east forming a subduction zone along the eastern edge of the plate (**Figure 13**). The Greater Antilles uplifted along the northern edge of the basin while Middle America formed along the south margin of the basin. The Cuban Arc collided with the Bahamas Bank between the Paleocene and Middle Eocene Epochs.

Between the Late Cretaceous Period and Middle Eocene Epoch, the spreading center between North and South America went extinct. The Laramide Orogeny uplifted the North and South American Cordillera (mountain belt) and within the Greater Antilles of the Caribbean.

The Caribbean Plate was inserted into the proto-Caribbean seaway between North and South America. Left lateral slip faulting (horizontal faulting), along transform fault and fracture zones of the Central Atlantic ridge system, extended into the Equatorial Atlantic Ocean (**Figure 13**).

The Caribbean Plate continued migrating eastward (**Figure 14**). South America migrated north towards North America, squeezing the Caribbean between the continents. Since Eocene time, NE convergence and eastward migration of the Caribbean plate with respect to North America occurred.

Since the Middle Eocene Epoch, slow but steady convergence occurred between the Americas resulting from strain collected by the western Atlantic Ocean fracture zones. Slow convergence suggests there were no significant plate boundaries in existence after 84 million years ago when the proto-Caribbean ocean basin formed.

West of Guatamala, the trench-trench-transform triple junction migrated to the SE, exposing the Mexican continental crust to uplift. Folding and faulting along the Middle American Trench occurred.

Structural fault and fold belts were cut off by the trench along the SW Mexican margin (**Figure 14**).

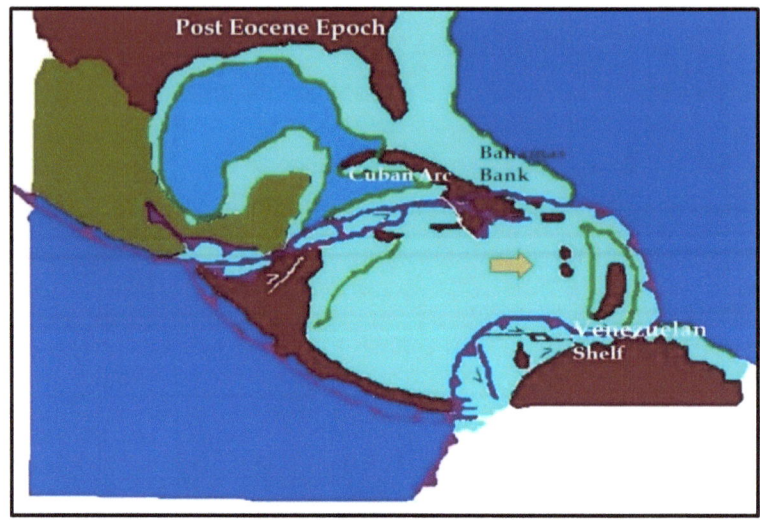

Figure 15. Island arc collisions.

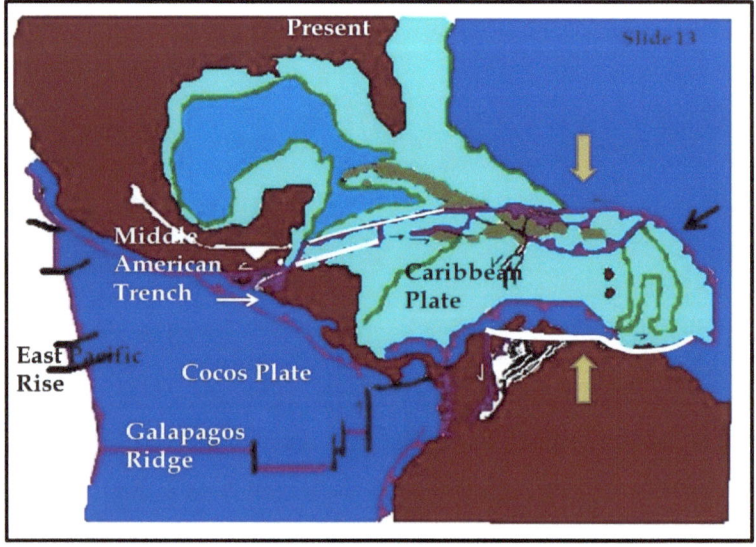

Figure 16.Caribbean plate compression.

Island arcs positioned along the southern Caribbean plate margin collided with Venezuela along the northwest South American margin (**Figure 15**). Northern South America and the Bahamas were stable shelves up until the Oligocene Epoch.

Accretion onto the Venezuelan Shelf formed nappes throughout the Caribbean mountain belt (**Figure 16**). Nappes are large overturned folds that were faulted and displaced by more than one mile from its original position. The Caribbean plate advanced, migrating eastward between the Americas.

South America continued squeezing the Caribbean Plate as it drifted northward towards the North American Plate. The Cocos Plate, Middle American Trench, East Pacific Rise, and Galapagos Ridge systems were well established. The Cocos Plate was well developed by the end of the Oligocene Epoch.

Northward convergence of the South American plate towards the North American plate started squeezing the Caribbean plate from the south towards the north. The subduction zone located off the northern coast of Venezuela transitioned into a right lateral slip fault along the eastern part of the zone. The eastern Caribbean subduction zone began consuming the South Atlantic plate's oceanic crust while the western part of the trench converted into a transform fault boundary.

The Cocos Plate was bounded by the Galapagos spreading ridge on the south, by the East Pacific Rise to the west, and by the Middle American Trench to the east (**Figure 16**).

c. The Nazca & Cocos Plates

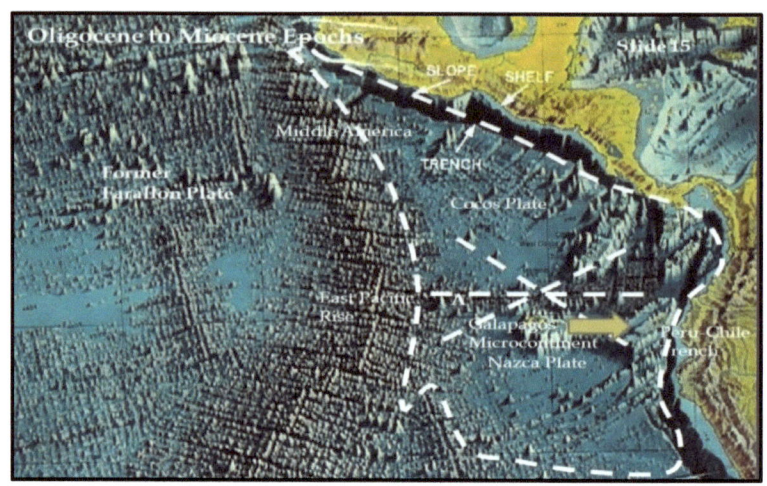

Figure 17. The Panama Basin .

Both Nazca and Cocos Plates make up the Panama Basin, subducting beneath the Panama Isthmus Middle America Trench (**Figure 17**). The East Pacific Rise provided the easterly drive for producing the Galapagas micro-continent while the Nazca-Cocos Spreading Center (A) provided the southward drive for rotating the Nazca Plate into the Peru-Chile Trench.

During the Late Oligocene Epoch, the Farallon Plate split into the Cocos and Nazca Plates. The plate split resulted in a global rearrangement of plate boundaries.

The oceanic crust of both Cocos and Nazca plates was formed along the Cocos-Nazca spreading center (A) and by the East Pacific Rise.

Figure 18.The Cocos Plate Ridge System.

The Farallon Plate split occurred between 22.8 and 19.5 million years ago. About 19.5 million years ago, the orientation of the spreading axis changed from NW to SE to an east-NE-west-SW direction up until 14.7 million years ago (white dashed lines, Figure 17). After 14.7 million years ago spreading was oriented from E to W. The Galapagos Micro-continent is presently being pushed towards the Peru-Chile Trench at right angles and will eventually accrete onto the continental margin (**Figure 17**).

The Cocos, Malpelo, and Carnegie Ridges (A) formed as the Farallon Plate passed over the Galapagos Hotspot (B) between 22 and 20 million years ago (**Figure 18**). South of Panama, the Colba Ridge was uplifted alongside a longitudinal transform fault (C), between the Late Miocene into Pliocene time. Hotspot volcanics covered oceanic crust formed at the three spreading systems which make up the Panama Basin (dashed line). The Cocos Ridge began to indent into South America about 2 million years ago (brown arrow).

D. The Mediterranean Sea

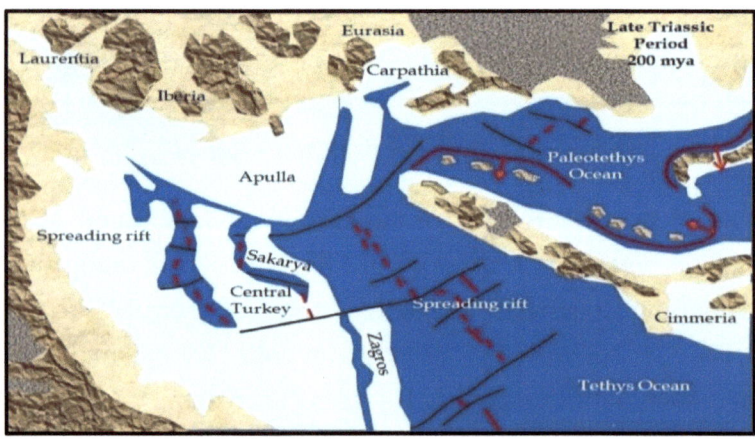

Figure 19. Closing of the paleo-Tethys Ocean, opening of the Tethys Ocean, and shallow sea invasion of the eastern African and southern Eurasian continents began during the Late Triassic Period.

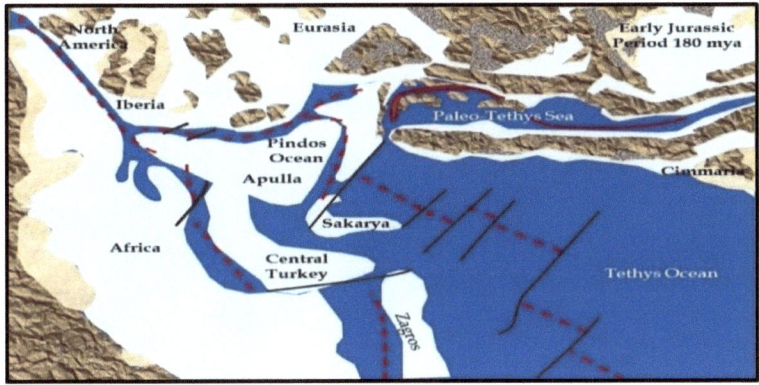

Figure 20. The breakup of Pangea rearranged plate boundaries by closing off the Paleo-Tethys Ocean. Cimmeria drifted closer to the southern Eurasian margin. Opening of the Tethys Ocean began along a spreading center positioned in the Tethys basin. Rifting allowed shallow seas to move into interior portions of central Eurasia. Shallow seas extended inland along eastern Africa.

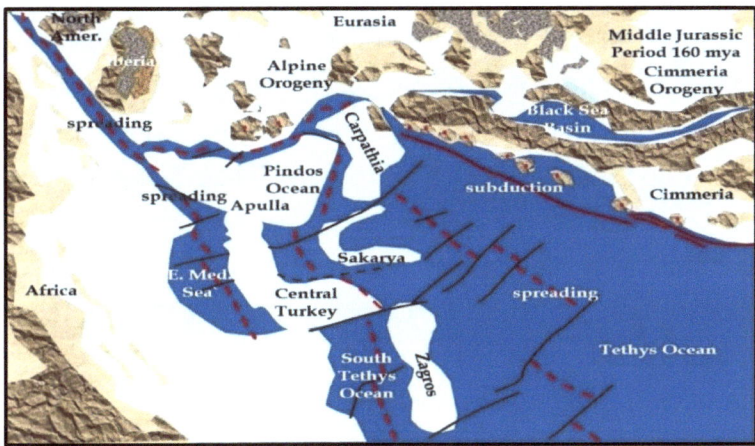

Figure 21. Cimmeria began to collide with Eurasia. The Tethys Ocean widened from an expanding spreading ridge center. Smaller ocean basins begin to develop along the eastern African margin. Shallow inland seas deepened within interior Eurasia, retreating in eastern Africa.

Figure 22. Inland seas receded from interior Eurasia, invading into eastern Africa. The Tethys spreading ridge system became inactive. The South Tethys Ocean activated along a new spreading ridge system. The smaller ocean basins off the eastern African margin coalesced into a discontinuous island arc system.

Opening of the Mediterranean Basin began with the closing of the Paleo-Tethys Ocean, and the drifting of Cimmeria towards southern Eurasia (**Figure 19**). Cimmeria separated the Tethys Ocean from the Paleo-Tethys Ocean. As Cimmeria approached Eurasia, the Paleo-Tethys Ocean became a shallow sea, connected to the main Tethys Ocean basin between Carpathia and the western end of Cimmeria. Several subduction zone segments and a spreading ridge were located within the Paleo-Tethys Ocean basin. Island arcs developed above the trenches. A mid oceanic ridge system occupied the main Tethys Ocean basin floor.

The Zagros (Iran) arc terrane occupied the western margin of the Tethys Ocean off the eastern African coast, covered in shallow seas (**Figure 20**). Sakarya and central Turkey were island arc systems separated from each other by spreading rifts in the shallow seas along the east and west margins of central Turkey. Laurentia was attached to western Eurasia. Iberia (Spain) was sandwiched between the two continents. Apulla (Italy) occupied the southern Eurasian margin, also covered by shallow inland seas.

Seas invaded as Pangea began to break apart. Shallow inland seas expanded covering much of the Eurasian lowlands and the eastern African margin. The paleo-Tethys Ocean began closing as Cimmeria drifted closer to the southern Eurasian margin. Volcanic arcs collided with the Carpathian Mountains above the western trench segment.

Cimmeria approached the eastern trench segment. The channel connection between the paleo-Tethys and Tethys Oceans was closed off.

Widening of the Tethys Ocean basin spreading center pushed Cimmeria towards the north. Off the eastern African coast, a new ocean floor began to develop the South Tethys Ocean. Zagros was pushed eastward by this new spreading center.

Sakarya and central Turkey began rotating counter clockwise along a broken transform fault segment located north of Zagros. Sakarya, central Turkey, and Zagros separated from the east African margin.

The Pindos Ocean began widening from a spreading center which expanded northward in the ocean basin. North America began to separate from Eurasia, and Iberia remained attached to Eurasia (**Figure 20**).

Inland seas continued to occupy continental lowlands (**Figure 21**). When Cimmeria collided with Eurasia, the Paleo-Tethys Ocean closed. The former Paleo-Tethys ocean basin formed the Black Sea Basin. East of the Black Sea, a mountain belt uplifted along the Eurasian margin. Continued spreading of the Tethys Ocean along the mid oceanic ridge system separated the main ocean floor basin from the South Tethys Ocean. Zagros drifted away from mainland Africa.

Sakarya separated from East Africa and from central Turkey along transform faults. A spreading center developed within the intervening basin. As central Turkey moved away from Africa, the Eastern Mediterranean Sea began to open along the basin spreading center.

The Pindos Ocean widened along a spreading rift. Carpathia drifted east from Apulla. The Alpine Orogeny began to develop north of Carpathia.

North America began to separate from Africa along a rift zone. Iberia remained wedged between North America and Eurasia. The paleo-Tethys subduction trench jumped to the south side of Cimmeria and a volcanic arc system developed in the fore arc basin (**Figure 21**).

As Eurasia and Africa began to rotate away from the South Polar Region inland shallow seas began to recede (**Figure 22**). Cimmeria accreted onto southern Eurasia closing the Black Sea Basin. The Cimmerian volcanic arc continued erupting along a portion of the continental margin as the old oceanic plate continued descending beneath the margin. Mountains were uplifted inland.

The Tethys Ocean spreading ridge center went extinct. The ocean basin stopped widening. The South Tethys Ocean spreading ridge center developed. Zagros continued to move away from eastern Africa.

Spreading between the African continent and central Turkey opened up the Eastern Mediterranean Sea. The Mediterranean spreading center developed. A subduction trench formed west of Turkey uplifting mountains along the western margin. Sakarya drifted closer to the accreted western Cimmeria terrane. Carpathia slid west along a transform fault closing the Pindos Ocean. Southern Eurasia pulled away from the Apulla terrane, opening up the Alpine Ocean.

Rifting between Iberia and Eurasia began forming. Iberia started to pull away from the western continental margin. The southern Cimmerian subduction zone broke apart into fragments. The trench jumped south to an off shore position (**Figure 22**).

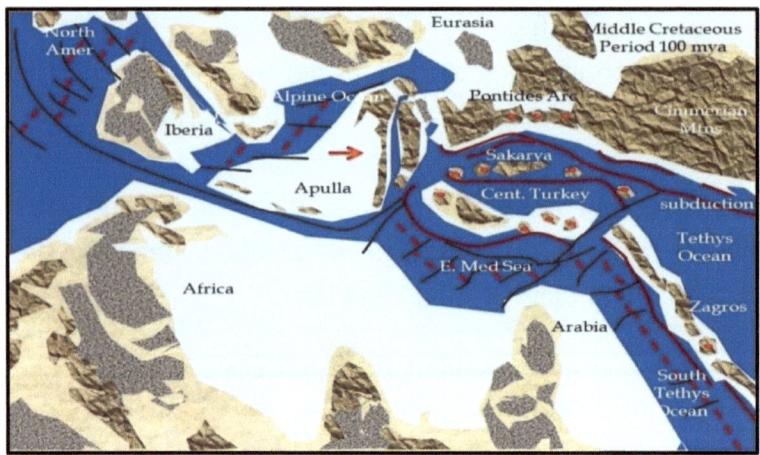

Figure 23. Cimmeria collided with the Russian Caucasus closing the paleo-Tethys Ocean. The Tethys Ocean basin narrowed. The South Tethys Ocean formed the Eastern Mediterranean Basin. Shallow inland seas expanded into northern Africa from eastern Africa. Seas began to retreat from interior Eurasia in response to the Cimmerian-Caucasus uplift. The island arc basins began to break apart by transform faults expanding from the South Tethys spreading ridge.

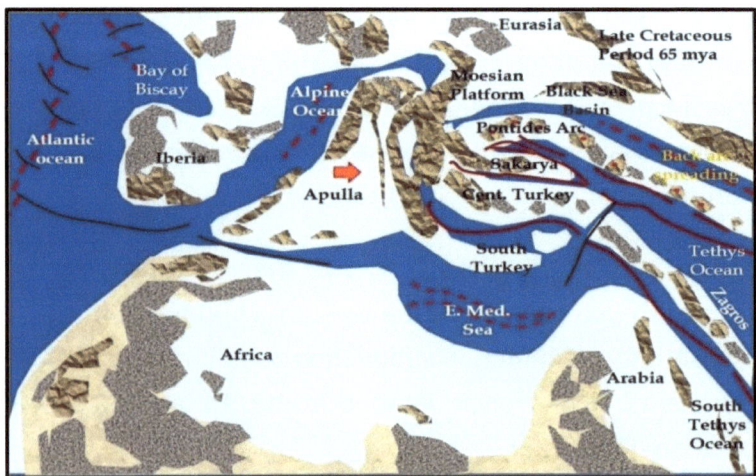

Figure 24. Island arc development in the Tethys Ocean began approaching the Eurasian margin compressing and uplifting linear mountain belts by rotation. Inland seas advanced again in Eurasia and northern Africa.

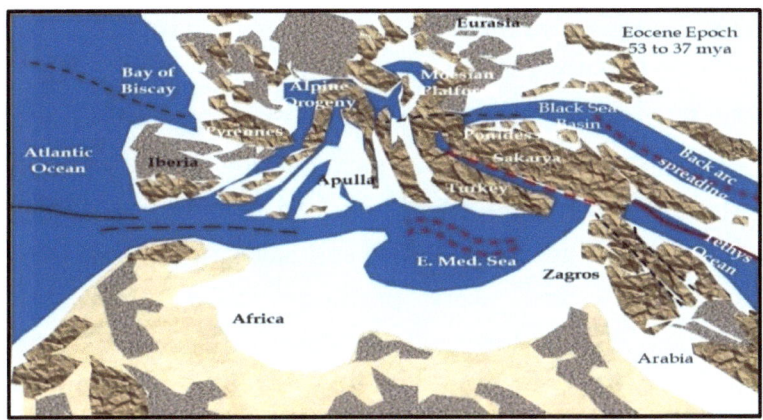

Figure 25. As the island arc system began to collide with the Caucasus region, the Zagros Mountains of Iran were uplifted. A back arc spreading center developed along the Russian margin. The Tethys Ocean was closing. The Eastern Mediterranean basin opened to the west connecting with the Atlantic Ocean. Shallow inland seas started to withdraw in the southern European region but remained stable in the Caspian Sea region of Russia. Seas began to withdraw from northern Africa.

Figure 26. Eurasian seas withdrew as the island arcs compressed against each other to form a continuous mountain belt from the Alpine Mountains southeast through the Zagros Mountain region. The western Mediterranean basin began to close off from the Atlantic Ocean. The Atlas Mountains uplifted along the northwestern Africa margin.

The Tethys Ocean basin was closing (**Figure 23**). The spreading centers in the South Tethys Ocean and Eastern Mediterranean Sea pushed Zagros, central Turkey, and Sakarya closer to south Eurasia. Volcanic arcs continued erupting above the Mediterranean and South Tethys trenches within central Turkey, Sakarya, and Zagros. Subduction zones were positioned along the south Eurasian margin forming the Pontides Arc above the trench. The Eastern Mediterranean Sea continued opening along the Mediterranean spreading center. A subduction zone formed along the south central Turkey margin. Volcanic eruptions and mountains were uplifted.

Eastern Apulla was being pushed up against the western edge of Carpathia. Mountain belts were thrust upwards by plate convergence. The Pindos Ocean closed. The Alpine Ocean began to extend to the west connecting with the widening Atlantic Ocean. North America continued to separate from Eurasia. Iberia rifted away from Eurasia along a spreading center separating Iberia from Eurasia (**Figure 23**).

A new back arc spreading center squeezed the Tethys Ocean from the north direction. Sections of the former Cimmerian terrane rifted away from south Eurasia (**Figure 24**). A subduction trench formed along the western Zagros margin. Consumption of the South Tethys Sea floor beneath Zagros, caused the ocean basin to close.

68

As the Pontides Arc erupted along the southern Black Sea basin margin, the Black Sea Basin re-connected with the back arc spreading center to the east. Sakarya began to collide with the western Pontides Arc and the Moesian Platform. Central Turkey edged closer to Sakarya. Along the northern margin of the East Mediterranean Sea, the South Turkey island arc began to develop. The Mediterranean Sea widened along a weakening spreading ridge.

The Azores Fracture zone extended from the Atlantic into the western Mediterranean between Apulla and Africa. The Alpine Ocean widened along a spreading center located west of Apulla, pushing Apulla against central Turkey and Sakarya. Thrusting and folding continued along the convergence zone to the east. The Atlantic Ocean continued to open. Iberia collided with Eurasia uplifting the Pyrennes along a suture zone. The Bay of Biscay opened along the northern Iberian coast (**Figure 24**).

As the Tethys Ocean was closing, the Zagros continent expanded in area by mountain belt collisions overriding island arcs (**Figure 25**). The back arc spreading center to the north continued to widen while the Black Sea Basin remained connected to the back arc basin to the east. To the west the basin was closed off by mountain belt collision and overthrusting. Sakarya and the Pontides Arc collided and accreted onto Eurasia uplifting a mountain belt which closed the western Black Sea margin. Central and South Turkey collided to form the Turkey mainland.

Turkey was moving closer to Sakarya. Apulla pushed closer to Turkey closing off the eastern Alpine Ocean.

Counter clockwise rotation of Apulla against the convergent thrust fold belt along the eastern margin formed the Hellenic Mountains (Greece). Rotation began to close off the northern Alpine Ocean basin. Apulla began colliding against southern Eurasia, initiating the Alpine Mountain uplift. Development of the Azores Transform Fault connected the Atlantic Ocean to the Mediterranean Sea (**Figure 25**).

Shallow inland seas withdrew from most of Eurasia (**Figure 26**). Seas continued to occupy the Caspian Basin and northern Africa. Collisions between Zagros and the Caucasus Mountains trapped the Tethys Ocean, creating a narrow inland sea. Both Caspian and Black Seas were closed basins cut off by the mountain ranges. Zagros continued converging together along thrust faults. Island arc and mountains were pushed on top of each other.

Africa continued migrating north towards the southern Eurasian margin. South Turkey and Central Turkey were sutured against Sakarya. Collision of the entire land mass was pushed against Moesia, building out the Eurasian continental margin to the south.

Collisions between Apulla and southern Europe closed off the Eastern Mediterranean Sea basin. Apulla began to assemble, and the Appenine Mountains of Italy uplifted. Evaporation of the Mediterranean Sea began.

As Africa approached Eurasia, a convergent thrust fold belt cut off the Eastern Mediterranean Sea from the Western Mediterranean Sea. The Atlas Mountain Belt was uplifted. As Apulla collided together, thrust faulting pushed the Alpine Mountains against Eurasia. Southern Europe began to look like it is today (**Figure 26**).

Figure 27. The Caspian inland sea retreated forming the Caspian Basin along the eastern edge of the Caucasus Mountains. The Black Sea basin was isolated. Shallow inland seas withdrew from southern Europe. The Mediterranean Sea was connected to a single basin with a narrow passage connected to the Atlantic Ocean on the west end near Gibraltar.

Collisions between Zagros and Eurasia closed off the Caspian Basin. Inland seas retreated, as the Caucasus Mountains were uplifted to the west of the Caspian Basin (**Figure 27**). Moesia was flooded along with southern Eurasia.

71

Figure 28. The Aegean Sea formed as the Africa plate began to subduct beneath southern Europe (dashed line). Subduction opened up a rift between Turkey and Greece, forming the Aegean Sea.

Convergent thrusting of mountain belts continued uplifting the Zagros Mountains. Arabia started to rift from Africa forming the Red Sea. Turkey closed off the southern Black Sea Basin and the Moesian Platform sealed the western basin margin.

Apulla pushed the Alpine Mountains on top of Eurasia, uplifting the Carpathian Mountains north of the Moesian Platform. This collision formed a convergent thrust belt along the suture zone between the Italian Alps and Europe.

Uplifting of the Hellenic Mountains formed Greece as a result of the Apulla thrust belt pushing east and north against Eurasia, Turkey, and Sakarya. The East Mediterranean Sea remained closed off from the West Mediterranean Sea by the southern tip of Apulla and the Atlas Mountains of northwest Africa. The West Mediterranean Sea was cut off from the Atlantic Ocean. Africa converged northward against Iberia. The Pyrennes continued uplifting along the suture zone between Iberia and France (**Figure 27**).

The Aegean Sea formed by crustal extension. African Plate subduction began beneath the Hellenic Arc (**Figure 28**). Normal faulting in the Late Miocene Epoch fragmented land masses, as subsidence affected the region.

Counter clockwise rotation produced transcurrent faulting in the Northern Aegean. This rotation was the last regional extension that occurred during the Late Miocene Epoch and Quaternary Period.

Rifting pulled apart the Hellenic Arc by north to south extension while compression occurred simultaneously in an east to west direction. Thrust faulting was caused by the Mediterranean Ridge being pulled against the Hellenic subduction trench (**Figure 28**).

E.   The Pacific Ocean

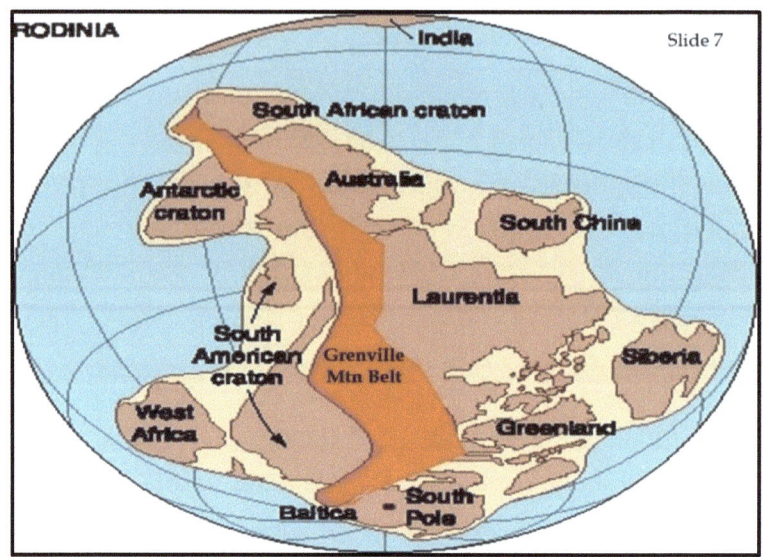

Figure 29. Rodinia Supercontinent.

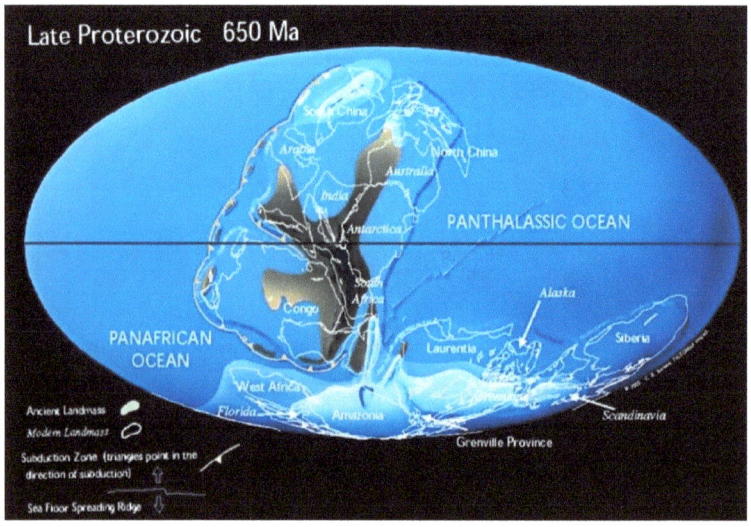

Figure 30. The Panthalassic Ocean.

74

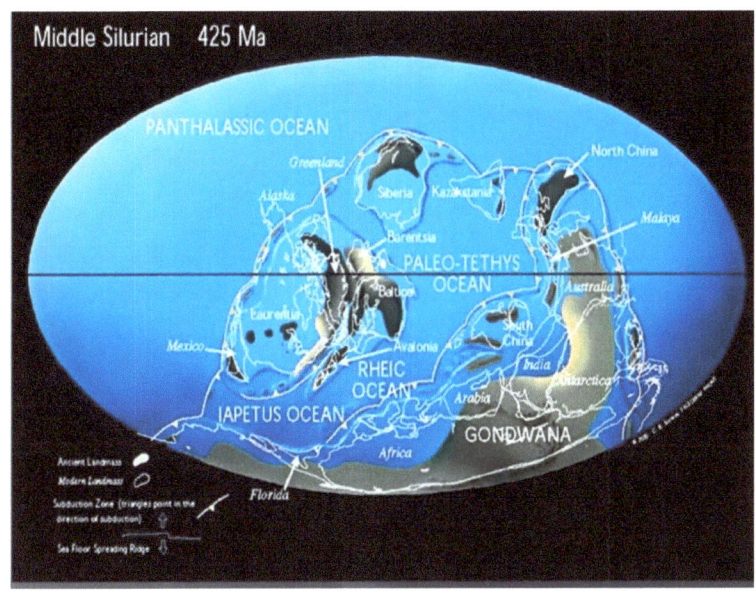

Figure 31. Many small oceans dotted the globe.

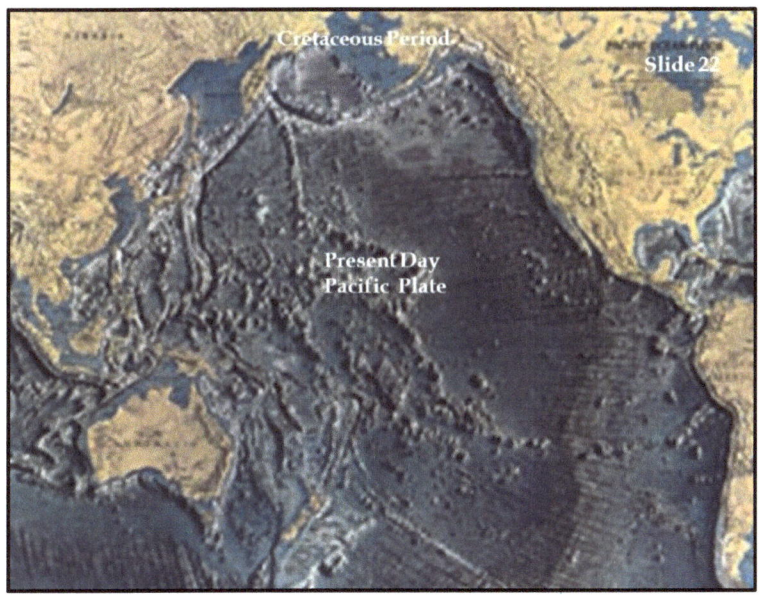

Figure 32. Pacific Ocean sea floor formed.

Rodinia was in existence between 760 and 726 million years ago (**Figure 29**). The supercontinent began to rift apart around 750 million years ago forming the Panthalassic Ocean.

Baltica was positioned at the South Pole. Greenland, South America, and Laurentia were positioned between 45 and 60 degrees south latitude. Siberia was located east of Laurentia and West Africa was located west of South America. South China, Australia, South Africa, and Antarctica were positioned near the Equator. India was located in the northern hemisphere.

South America was sutured to Laurentia by the Grenville Mountains (rust red, Figure 29). The Grenville Province stretched from Baltica at the South Pole north to the South African craton.

The opening of the Panthalassic Ocean was triggered by a super plume positioned along a spreading center to the east of the supercontinent (**Figure 30**). The first spreading centers began the process of continental drift. During the Cambrian Period most of the continents positioned at the South Pole were covered in glacial ice including portions of Laurentia, Baltica (Scandinavia), South America (Amazonia), West Africa, and Florida.

The exposed continental mass formed a linear belt spreading in a dendritic pattern to the north, northwest, and west directions from approximately 50 degrees south latitude through the Equator to around 50 degrees north latitude. Exposed rocks observed in South Africa, the Congo, Antarctica, India, Arabia, and Australia were used to assemble the former continent together. South China and Australia were positioned near the Arctic Circle, partially covered by glaciers. A plate boundary occupied the continental land mass on all sides. Exposed rocks support the pattern shown in **Figure 30**.

The Panthalassic Ocean covered most of the eastern hemisphere. A spreading center was positioned in the eastern ocean, trending northeast from the Southern to the Northern Hemisphere. A possible subduction zone was located along the eastern plate margin.

A subduction zone rimmed the western margin, extending from the Arctic Circle southward to around 60 degrees south latitude. The Panthalassic Ocean occupied most of the northern hemisphere while smaller ocean basins occupied the area between Laurentia and Gondwana. Small island arc systems developed above subduction zone complexes. These arc systems eventually collided with larger land masses. The continents began to build out (**Figure 30**).

About 400 million years ago Pangea began to form, centered on the Equator (**Figure 31**). Baltica collided with Greenland and eastern Laurentia. The paleo-Tethys Ocean widened between Gondwana and Laurentia.

The Paleo-Tethys Sea was positioned to the northeast and the Panthalassic Ocean was positioned to the west of Laurentia. Several smaller plates had developed since Cambrian time.

The Laurentide and Siberian plates occupied the western paleo-Tethys Sea. Baltica and Greenland were positioned at the Equator. North China, Malaya, Australia, Antarctica, India, Arabia, and Gondwana occupied the eastern Paleo-Tethys Sea. Subduction zones rimmed continental plates. Island arc systems developed above most of the subduction zones.

Avalonia, Mexico, and a set of arcs were positioned in the western Rheic Ocean and northern Iaepetus Ocean. Intra-plate arcs developed over hot spots within the western Laurentian continent. Island arcs developed east of Siberia, Kazakhstan and off the western Gondwana margin (**Figure 31**). The Panthalassic Ocean occupied the surrounding global area.

The Pacific Ocean formed as sea floor spreading began to push new oceanic crust outward onto the Pacific basin floor during the Jurassic Period, 167 million years ago (**Figure 32**).

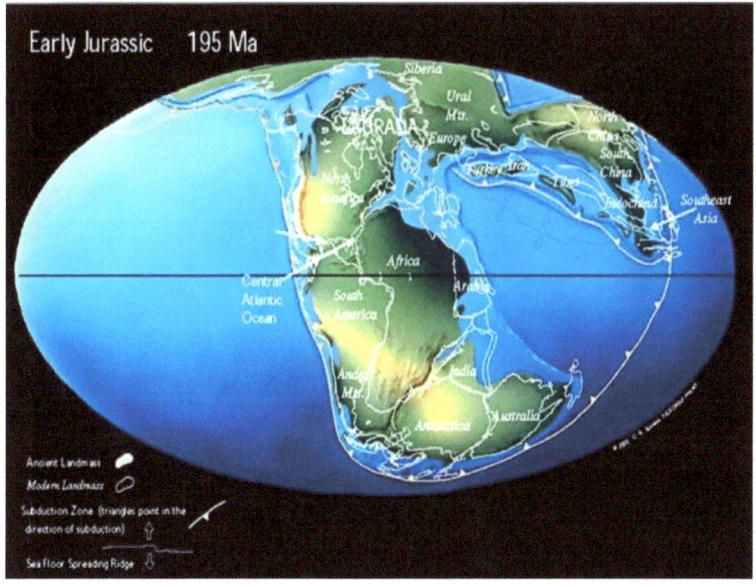

Figure 33. Laurentia-Gondwana collision.

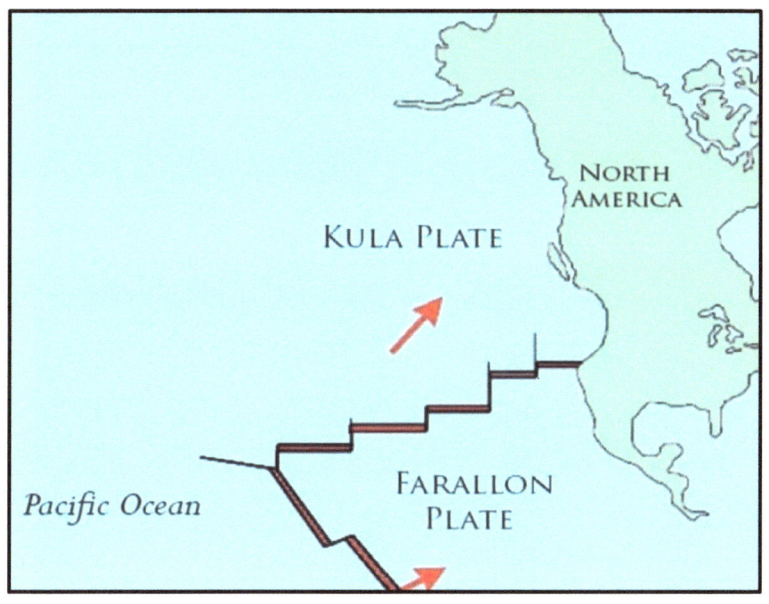

Figure 34. Farallon Plate splitting.

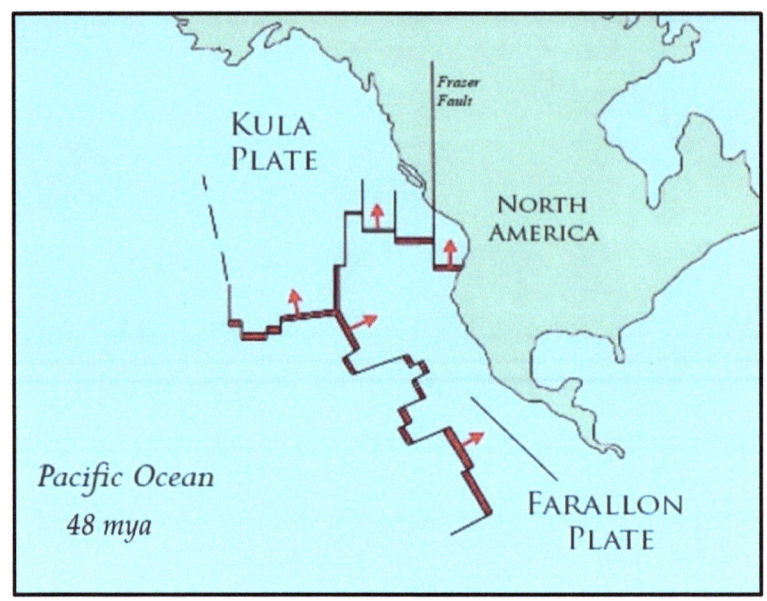

Figure 35. Kula Plate shift.

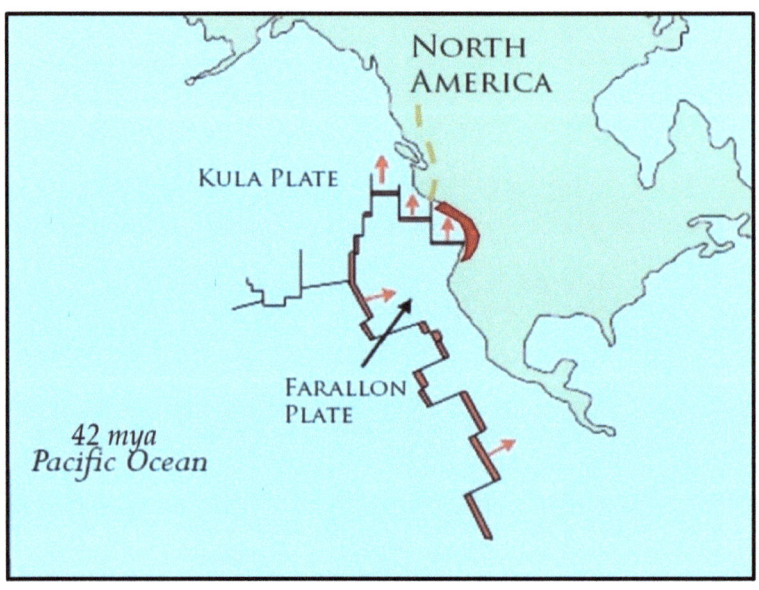

Figure 36. Fraser fault development.

A new sea floor plate formed in the Pacific Basin, called the Farallon Plate. About 140 million years ago, the Pacific Plate was subducting beneath South America (**Figure 33**). About 120 million years ago, the Farallon Plate was split by a new spreading ridge. The new ridge began the splitting of the Kula Plate from the Farallon Plate, towards the north.

Kula Plate motion was directed to the northeast off the Pacific Northwest coast (**Figure 34**). A shift in direction occurred to the north, as subduction transitioned into a transform fault boundary along the western margin, 58 million years ago. The Kula-Farallon ridge was oriented along an east-west axis south of the Columbia Embayment.

Between 58 and 38 million years ago, the Kula Plate broke apart along a series of north-south transform faults (**Figure 35**). A transform boundary created the Fraser Fault along the north trending Kula spreading center. The fault separated the Kula Plate into an eastern and western segment. As the plates split apart, the eastern segment was subducted beneath the Columbia Embayment, erupting Challis volcanics from a new magmatic arc system. Magma was routed along the western segment, flowing out onto the ocean floor along spreading ridge transform faults.

The Fraser Fault changed displacement along the continental margin from a transform boundary to a lateral strike slip displacement. Fault displacement ended the Kula-Farallon Ridge spreading.

Challis magmatic activity ended when the Farallon Plate replaced the Kula Plate. Plate shifting to the north helped extinguish Challis arc activity (**Figure 36**).

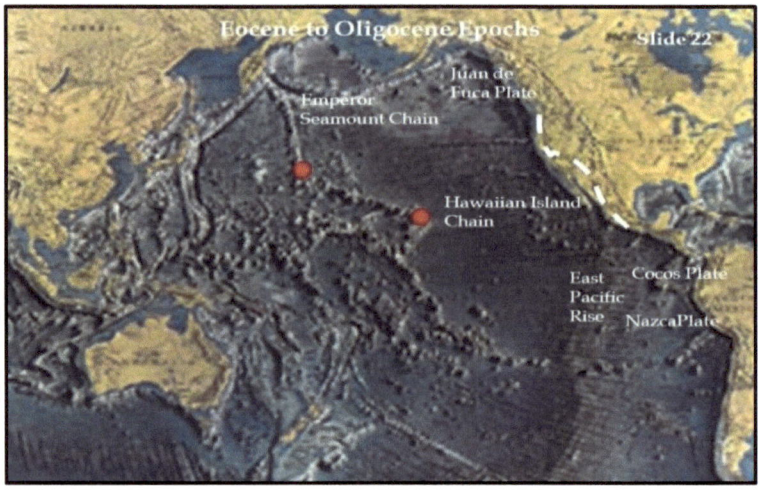

Figure 37. Emperor Seamount Chain kink.

Consumption of the Kula Plate along the Alaskan Aleutian Island chain gave rise to the Pacific Plate, 33 million years ago, covering two thirds of the Pacific Ocean. The East Pacific Rise separated the Pacific Plate from the Farallon Plate. Collision between the Farallon Plate spreading ridge center with California, split the plate into the Juan de Fuca Plate to the north and the Cocos-Nazca Plates to the south (**Figure 37**).

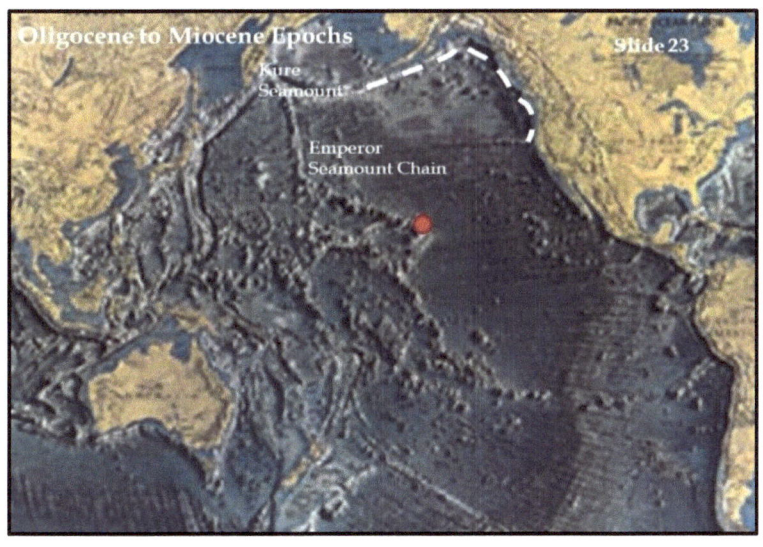

Figure 38. Kula Plate consumption.

Figure 39. Hawaiian Hot spot.

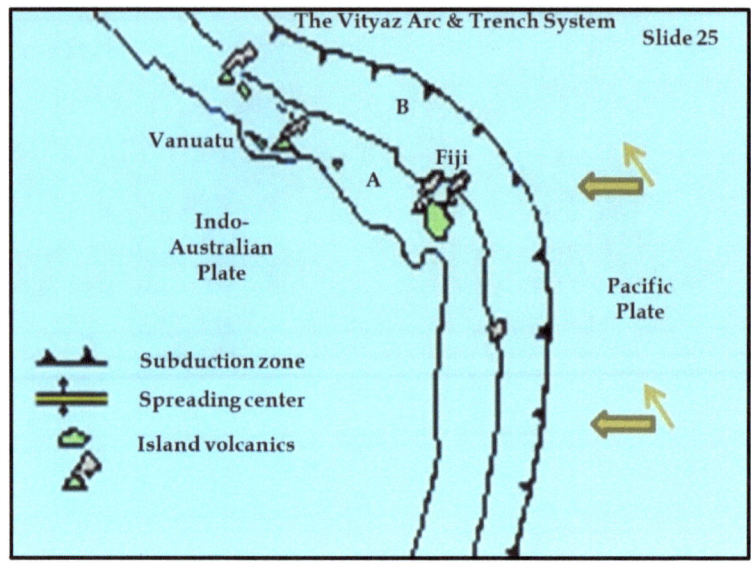

Figure 40. Vityaz Arc & Trench System.

Seamounts represent old volcanoes where the upper portions of the volcanoes were later eroded by wave activity occurring over millions of years. The Emperor Seamount chain was formed by the same hot spot that is currently forming the Hawaiian Island volcanic eruptions. Positioned at the northern end of the seamount chain is the Kure Seamount, thought to be about 28 million years old. The Island of Hawaii is about 400,000 years old. The Kula Plate was completely consumed beneath Alaska ending the plate's existence 20 million years ago (**Figure 38**).

The Hawaiian Island chain is the southernmost extension of the Emperor Seamounts which extend northwest into the Aleutian Trench (**Figure 39**).

The archipelago (or island arc) formed as the Pacific Plate moved northwest over a hot spot now located beneath the Island of Hawaii. Recent theories suggest the Hawaiian hot spot shifted in response to a change in heat flow circulation patterns within the solid upper mantle about 45 million years ago.

This hot spot shifting is now believed to be the cause of the kink between the Emperor Seamount chain and Hawaiian Volcanic chain. Eruptions consist of very fluid type lavas, producing basalt, diabase, and gabbros, unlike the explosive type andesitic eruptions common to the Pacific Ring of Fire (**Figure 39**).

Between 40 and 20 million years ago, a continuous island arc called the Vityaz Arc (A), formed by westward subduction of the Pacific Plate beneath the Indo-Australian Plate at the Vityaz Trench (B) (**Figure 40**). Between 20 and 12 million years ago, the Pacific Plate boundary changed direction from west to northwest. The change in direction shifted plate motion the same way along the northern Vityaz Trench. Subduction flattened out to a shallower angle northwest of Fiji. Little change occurred to the south of Fiji.

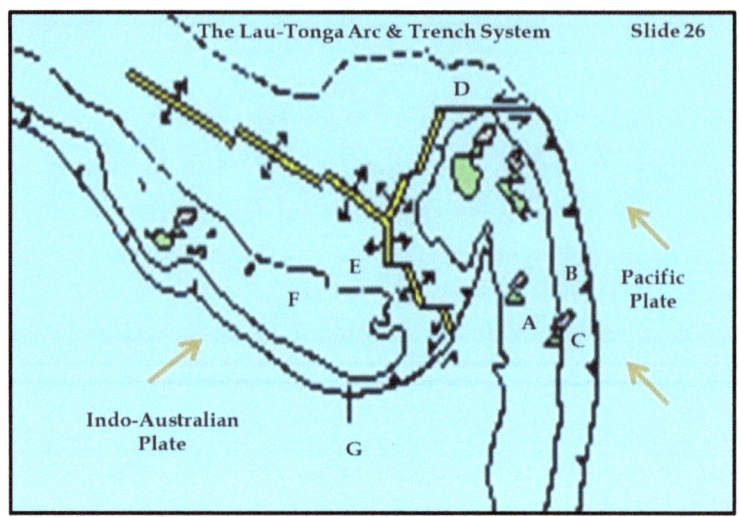

Figure 41. Lau - Tonga Arc.

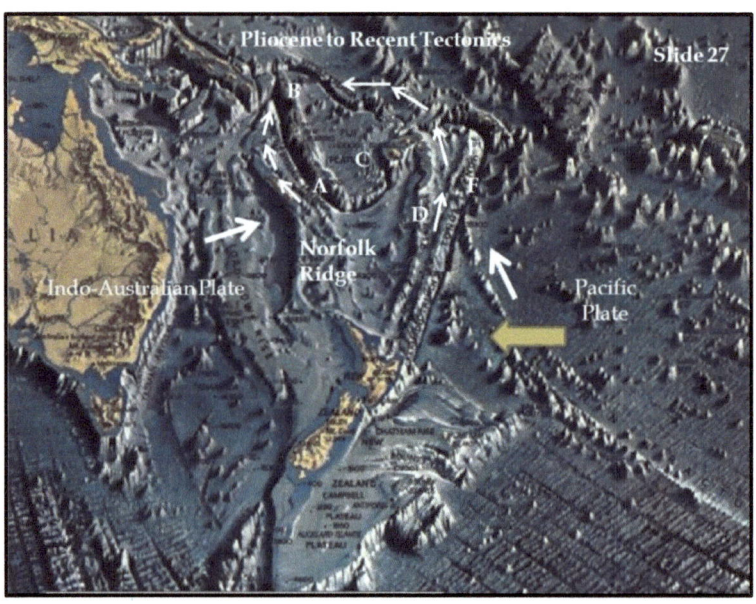

Figure 42. New Hebrides Arc.

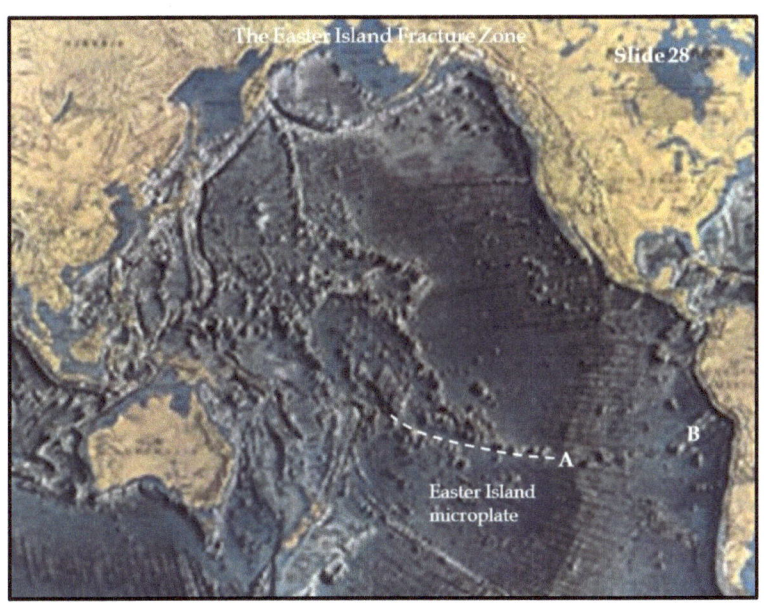

Figure 43. Easter Hot Spot.

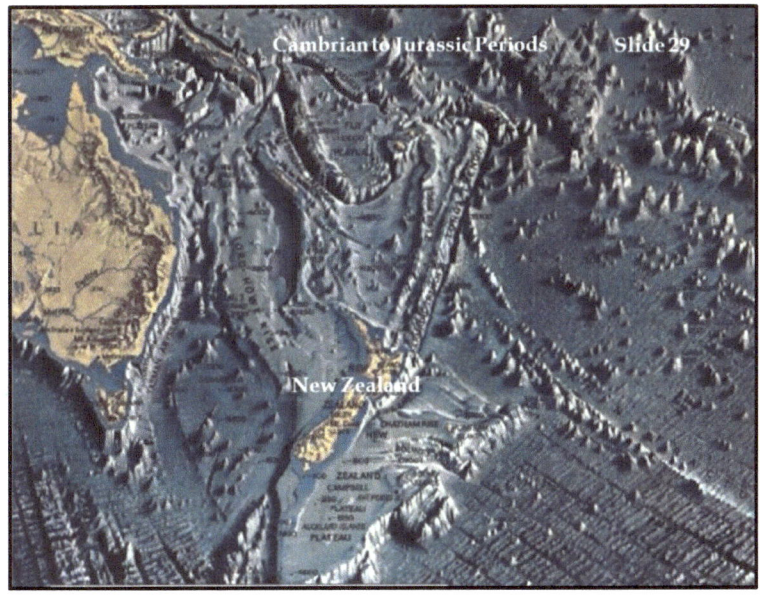

Figure 44. New Zealand.

87

A new volcanic island arc called the Lau-Tonga Arc (A) formed the Vitu Levu, Yasawa, and Mamanuca group of islands (A) (**Figure 41**). Continued volcanism occurred on the Lau-Tonga Arc. The Pacific Plate subducted northwest beneath the Indo-Australian plate along the Vityaz Trench (B). Plutonic rocks were injected into the island arc crust and earlier volcanic rocks were folded and faulted. Uplift occurred on Vitu Levu (C). From 12 to 7 million years ago, subduction continued northwest between the Pacific Plate and the Indo-Australian Plate. The Indo Australian plate moved to the northeast. The Vityaz Arc split along a lateral strike slip fault, called the Fiji Fracture Zone (D). Subduction ended at the Vityaz Trench, as the Vityaz Arc broke up (E), forming the Vanuatu Arc and trench system (F, G).

From 7 million years ago to the present, the New Hebrides Trench formed (A), as subduction of the Pacific Plate continued drifting to the northwest (**Figure 42**). The Indo-Australian plate began subducting beneath the Pacific Plate northeastward along the New Hebrides section of the former Vityaz Trench. Volcanism renewed along the New Hebrides Arc (B). The New Hebrides Arc began to rotate clockwise while Fiji rotated counter-clockwise away from the Vityaz Trench segment. Opposing rotations began sea floor spreading and the opening of the North Fiji Basin which became a back arc basin to New Hebrides (C). The Lau-Tonga portion of the former Vityaz Arc began splitting up.

Sea floor spreading accompanied by volcanism formed the Lau Arc (D) and the Tonga Arc (E) separated by the Lau Basin, 2.5 million years ago.

Sea floor spreading in the Lau Basin isolated the Lau Arc from subduction along the Tonga Trench (F). Active island arc volcanism is occurring on the Tofua Arc, west of the Tonga Arc, resulting from westward subduction of the Pacific Plate beneath the Indo-Australian Plate at the Tonga Trench.

The Easter Hot Spot, near Easter Island, formed as an intra-plate volcanic island, east of the East Pacific Rise (A) (**Figure 43**). About 4.5 million years ago, the Easter Island micro-plate was formed when the Easter Fracture Zone separated the Pacific Plate from the Nazca Plate. Island arc chains also formed on the Nazca Plate, called the San Felix and San Ambrosia chains off the Chilean coast (B).

Between 545 and 370 million years ago, New Zealand was part of Gondwana, undergoing sedimentation and volcanism (**Figure 44**). Between 370 and 330 million years ago, sea floor sediments were uplifted, folded, and metamorphosed during the Tuhua orogeny. Intrusions also occurred into the continental crust as New Zealand continued to remain a part of Gondwana. Between 330 and 142 million years ago, northwest New Zealand underwent a period of sedimentation filling up the New Zealand Geosyncline along the continental shelf and slope.

The Rangitata orogeny compressed the geosyncline sediments as they were folded and uplifted along with sea floor basalts in the South Island. Simple folding occurred in the west block while severe thrusting and folding occurred in the east block. A geosyncline is a linear trench located off a continental margin which receives sediment from both continental and oceanic crust. A geosyncline is the same as a subduction trench.

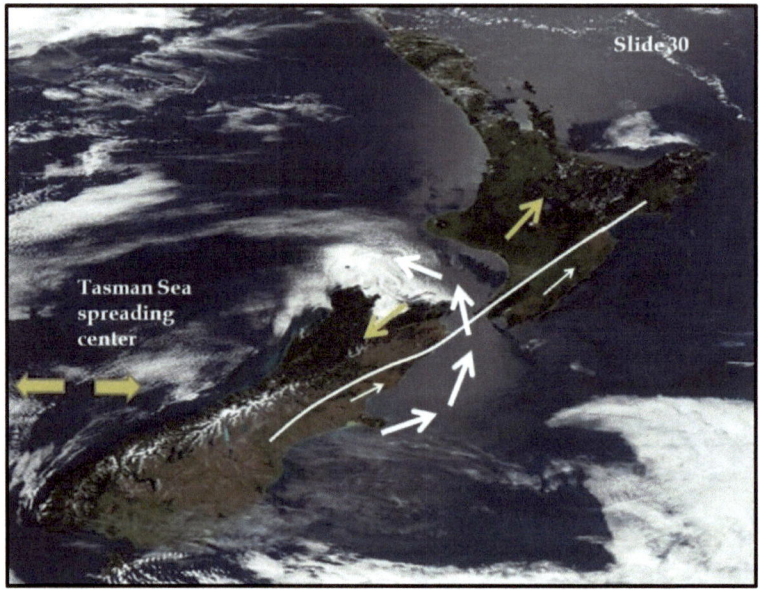

Figure 45. New Zealandia.

New Zealand began to separate from Gondwana about 90 million years ago (**Figure 45**). The southern part rotated counter-clockwise. Sea floor spreading followed between 83 and 62 million years ago, expanding north into the Coral Sea.

The northern parts separated when the Tasman Sea was created up until 52 million years ago when sea floor spreading ended. A new rift valley formed between Australia and Antarctica. Sea floor spreading continued, 24 million years ago. Strain built up in the SW Pacific Plate resulting in the formation of strike slip faulting.

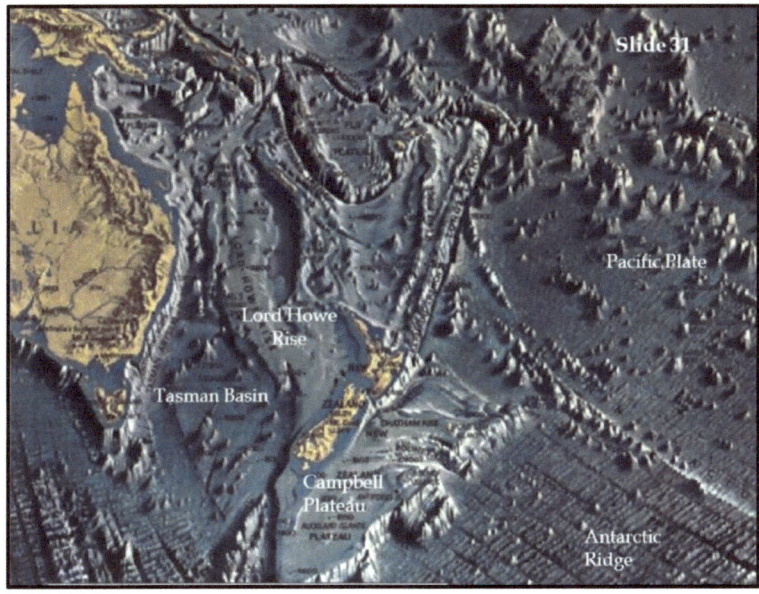

Figure 46. The Campbell Plateau.

Central Westland uplifted, producing the Southern Alps along the Alpine Fault. Between 20 and 10 million years ago, the west side moved northeast relative to the east side. Between 10 million years ago and the present, tectonism continued to uplift mountain ranges on both islands. Pacific Plate subduction caused volcanism on the North Island which gradually migrated south to eventually reach the Taupo Volcanic Zone (**Figure 45**).

The Campbell Plateau formed as part of Zealandia, after New Zealand rifted away from Gondwana, 90 million years ago (**Figure 46**). The plateau was submerged about 52 million years ago as part of sea advances across New Zealand. Stewart Island, the Aucklands, and Campbell Island still remain emerged on the plateau.

F. The Indo-Australian Plate

Figure 47. The Southeast Indian Ridge.

The Southeast Indian Ridge (A) is pushing the Indo-Australian Plate north towards the Indonesian Islands (B). The Australian Plate is being pulled into the Java Trench beneath Indonesia (C). To the west of Australia, basin ridges appear distorted (**Figure 47**).

Figure 48. The Broken Ridge.

Broken Ridge (A) is being distorted by squeezing occurring between the Southeast Indian Ridge, the northward motion of the Indo-Australian Plate, and the stuffing of the plate into the Java Trench. Australia may be rotating slightly to the right pushed by the opening of the smaller basin off the southeastern continental shelf (**Figure 48**).

Subduction beneath the Pacific Northern New Guinea Ridge (A) led to massive volcanic eruptions forming the Izu-Bonin-Mariana arc system (B), about 45 million years ago. (**Figure 49**)

G. The Philipine Plate

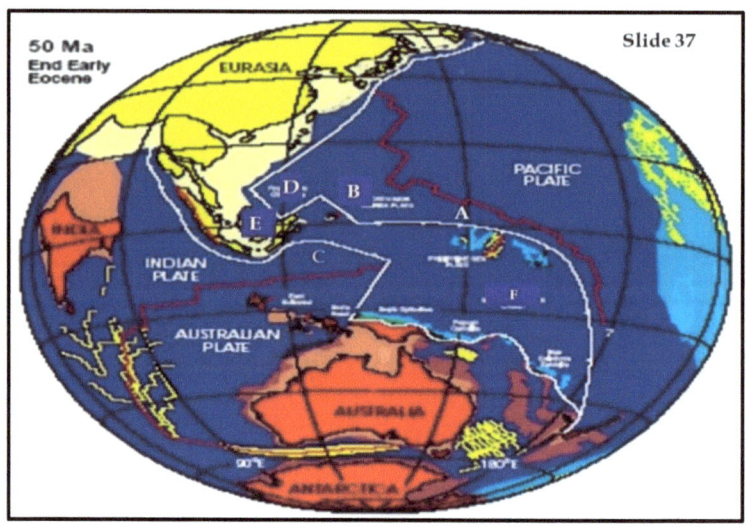

Figure 49. New Guinea Ridge.

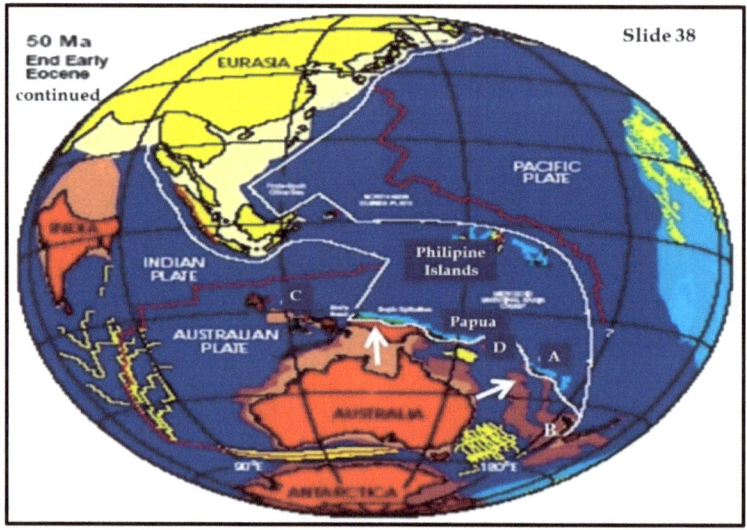

Figure 50. The Java-Sulawesi Trench System.

94

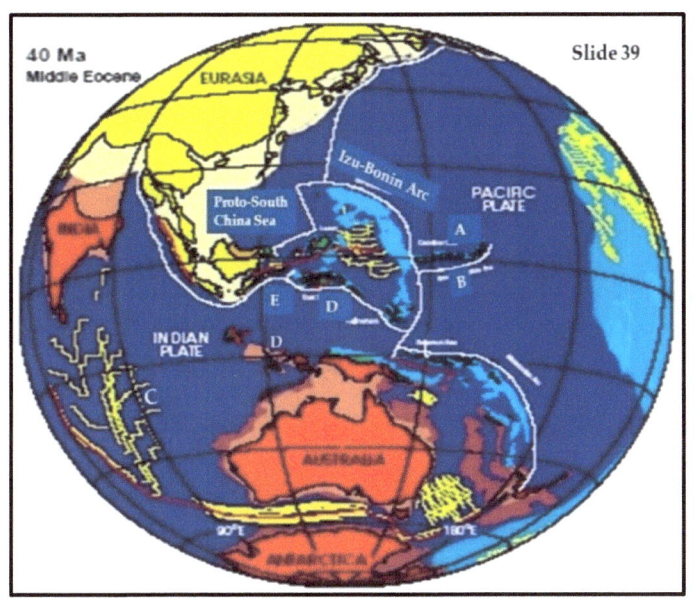

Figure 51. Caroline Sea Back Arc Spreading.

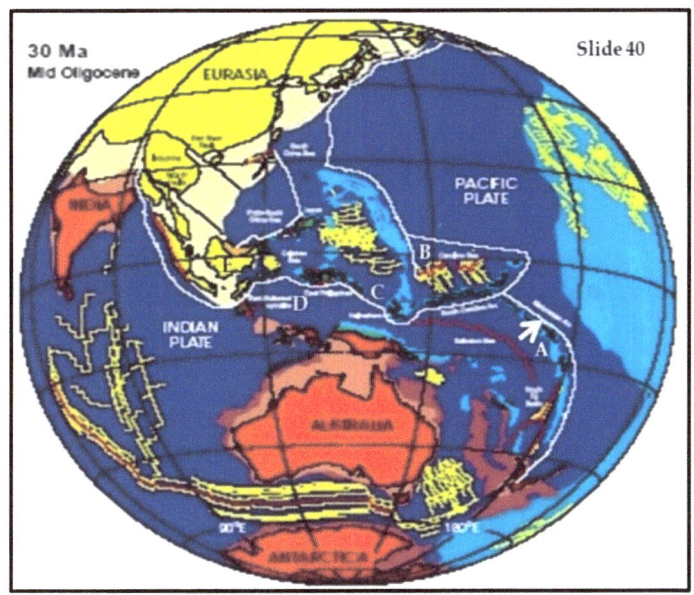

Figure 52. The Melanesian Arc.

When the west Philipine-Celebes Sea basin opened (C), southward subduction of the proto-South China Sea beneath Luzon and the Sulu arcs occurred (D).

Subduction of the proto-South China Sea pulled the basin apart along the basin margin. At the same time, the marginal slab was pulled downwards due to subduction between Borneo and Luzon (E). Sea floor spreading followed in the South China Sea. Opening of the South China Sea coincided with the creation of the Philipine Sea plate (F).

During the Eocene Epoch, the Australian passive margin collided with New Guinea to form New Caledonia (A) (**Figure 50**). Subduction beneath Papua-New Guinea formed New Britain, Solomon, and the Tonga-Kermadec arcs (B). Northward subduction of the Indian-Australian plate continued beneath the Sunda-Java-Sulawesi arcs (C) forming rift basins throughout Sundaland. The Java-Sulawesi trench extended into the West Pacific beneath the east Philipine and Halmahera arcs (D).

Back arc spreading of the Caroline Sea (A) started forming about 40 million years ago due to subduction of the Pacific Plate (**Figure 51**). The Caroline Arc formed at this time (B). The South Caroline arc became the north New Guinea arc terrane.

Spreading of the western Indian Ocean (C) and SW Pacific marginal basins, and the continued subduction of the Sunda-Java trenches through Sulawesi (D), the Philipines, and Halmahera (D) formed the Celebes Sea (E), lasting up to about 34 million years ago.

As the Solomon Sea was being destroyed by subduction (A), the Melanesian arc was migrating northward (**Figure 52**). The Caroline Sea was widening above a subduction zone (B). Philipine-Halmahera island arc positioning remained stationary at this time (C). Spreading of the Philipine-Celebes Sea (D) continued subducting between NE Borneo and north of Luzon (E).

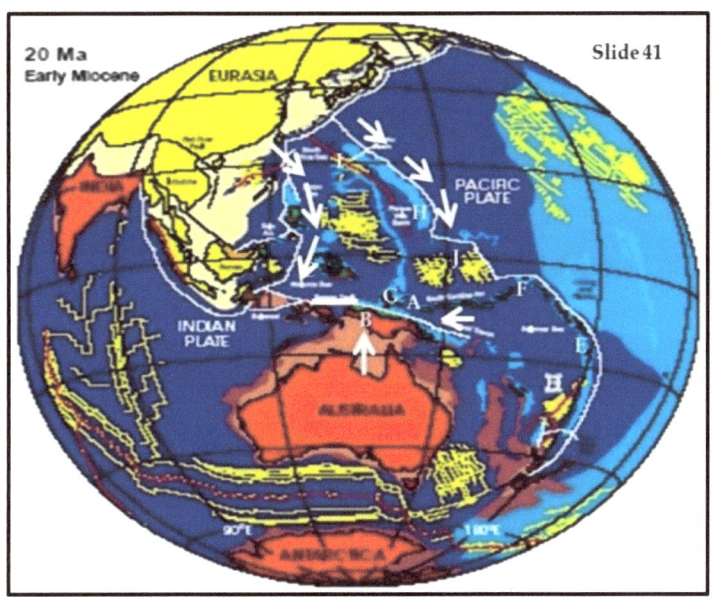

Figure 53. New Guinea collision.

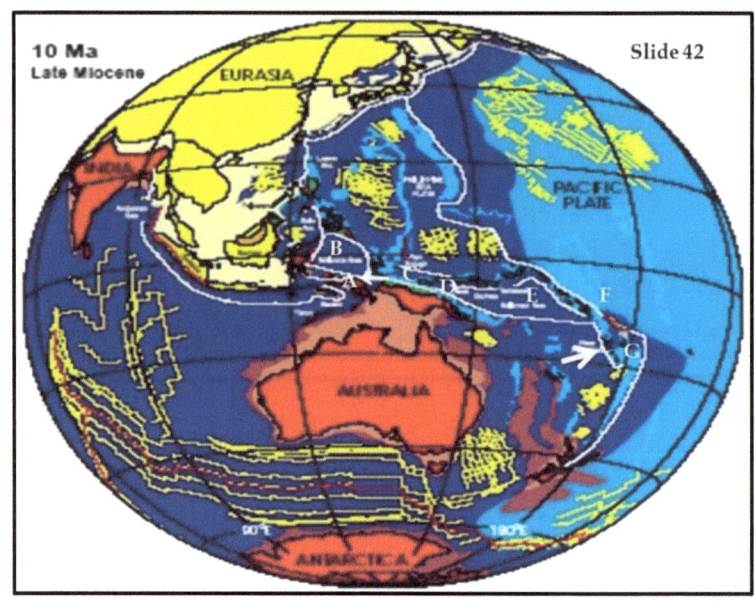

Figure 54. The Halmahera Arc formation.

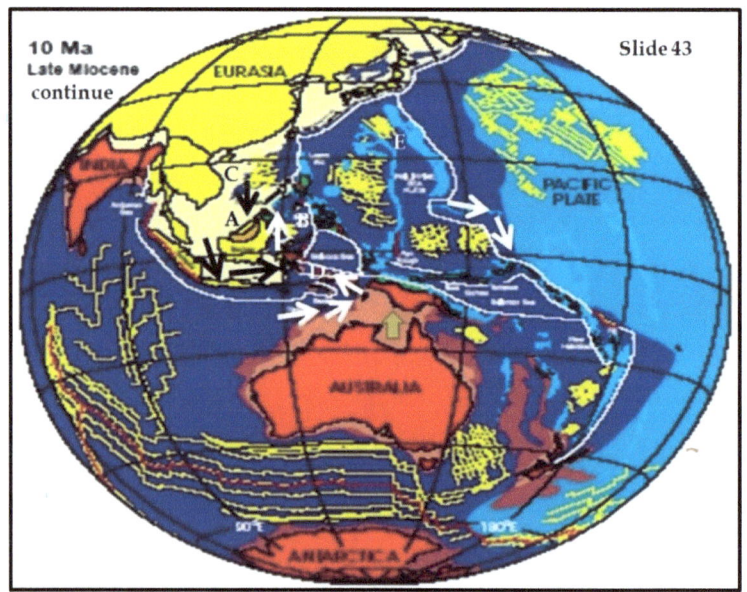

Figure 55. Borneo rotation.

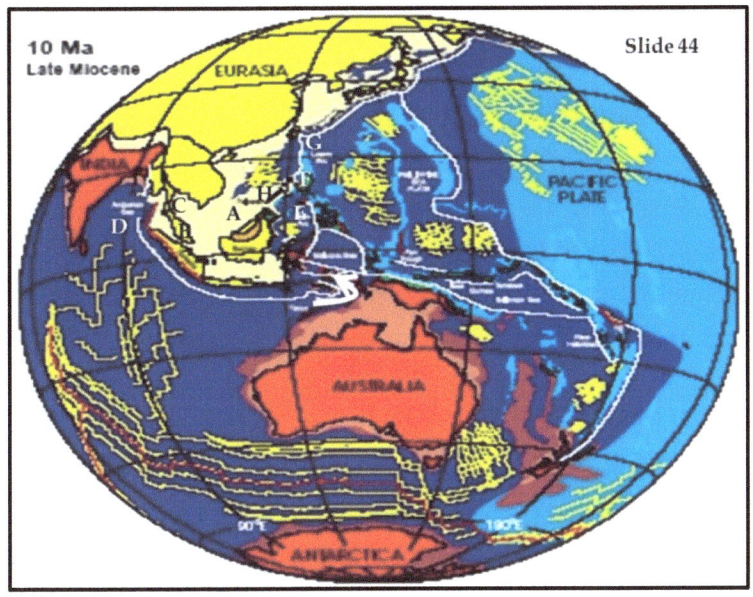

Figure 56. Rotation of Indochina & Indonesia.

About 25 million years ago, the New Guinea passive margin collided with the leading edge of the east Philipine-Halmahera-New Guinea arc system (A) (**Figure 53**). The Australian margin was approaching west Eurasia near West Sulawesi (B). When the Australian margin reached the Eurasian subduction zone northward subduction ended.

The trapped oceanic crust between Sulawesi and Halmahera became part of the Philipine Sea Plate and later on, the Molucca Plate (C). The Philipine Plate began rotating clockwise and the trapped oceanic crust began to subduct beneath Sulawesi to form the Sangihe arc.

Between 25 and 20 million years ago, the Ontang-Java platform (D) collided with the Melanesian arc (E) connecting Melanesia to New Guinea along the southern Caroline Plate margin (F) and to the Philipine-Halmahera arc (G). Subduction ended along the Philipine-Halmahera arc. The New Guinea segment became a strike slip fault (called the Sorong Fault), shifting the South Caroline arc along the New Guinea margin. Both Caroline and Philipine plates began to rotate and the Izu-Bonin-Mariana plate (H) migrated eastward into the Pacific. Rifting opened up the Parece Vela (I) and Shikoku basins (J). Subduction began along the Asian margin (**Figure 53**).

Accretion onto Sulawesi caused subduction to begin at the eastern margin of the Molucca Sea forming the Halmahera arc (A) (**Figure 54**). The Molucca Sea became a separate plate (B).

The opening of the Ayu Trench (C) separated the Caroline plate from the Philipine Sea plate. Westward motion of the Caroline arc along a strike slip fault resulted in accretion of the Caroline arc onto the New Guinea arc (D). When the Caroline arc collided with the New Guinea arc, the collision shifted subduction of the Solomon Sea plate from the south to the north beneath New Britain (E). Collisions between Java-Ontang with Melanesia caused the Solomon Islands to attach to the Pacific Plate (F).

Subduction to the west began on the SW side of the Solomon Sea beneath eastern New Guinea, eliminating most of the Solomon Sea. The Maramuni arc system formed. The Southern Solomon Sea subducted eastward creating the New Hebrides arc (G), ending the opening of the South Fiji basin. Most of the Solomon basin was destroyed.

Between 20 and 10 million years ago, Australia was moving north causing Borneo (A) to rotate counter-clockwise (**Figure 55**). Volcanism along the North Borneo margin shed sediments into the western proto-South China Sea (B). As the Philipine Sea plate rotated clockwise, the South China Sea spreading center re-oriented its position to form new subduction zones along the Eurasian eastern margin (B) and in the SW Pacific. The North China Sea continental margin crust thinned (C). The margin was then thrust beneath Borneo, causing the Borneo crust to thicken. Crustal thickening caused shedding of sediments into the Borneo continental shelf to form deltaic deposits.

North Sumatra rotated counter-clockwise along with South Malaya (D). Volcanism increased, weakening the upper plate forming the Sumatran strike slip fault. Subduction rates increased as a result of the rotation. The Philipine Islands continued to migrate on top of the Philipine sea plate, erupting volcanics. The east edge of the plate ended in the Shikoku basin (E) (**Figure 55**).

As Borneo rotated (A), west Sulawesi and Sundaland rotated counter-clockwise (**Figure 56**). Malay (B) rotated clockwise but remained attached to Indochina and to the south Malay Peninsula. The Gulf of Thailand opened up (C).

Burma became coupled to the Indian Plate moving north along the Sagaing Fault, stretching the Sunda continental margin north of Sumatra. Stretching of the Sunda continental margin opened the Andaman Sea (D). East of Borneo, increased subduction along Sundaland caused the Sulu Arc to split opening the Sulu Sea as a back arc basin south of the Cagayan Ridge (E). The ridge moved north, closing the proto-South China Sea. The ridge collided with the Palawan margin.

The Philipines and Halmahera migrated on the Philipine Sea Plate towards the Luzon subduction zone (F). Strike slip faulting connected the west margin of the Philipine Sea plate to the subduction zone at the Ryuku trench (G). Luzon & Cagayan collided with the Eurasian continental margin (H) in Mindoro and Palawan causing the trench to jump south of the Sulu Sea (**Figure 56**).

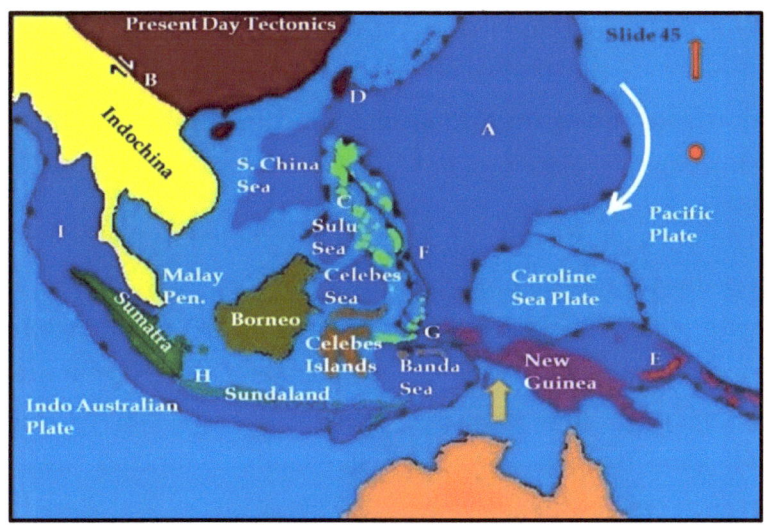

Figure 57. Philipine plate rotation.

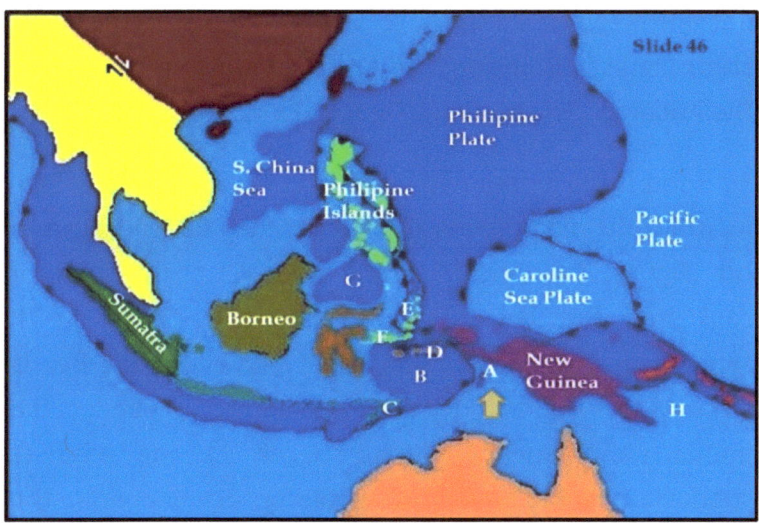

Figure 58. Banda Sea rifting.

About 5 million years ago, the Philipine Sea Plate (A) rotation pole shifted north from a former position east of the plate (**Figure 57**).

103

The plate continued to rotate as new plate boundaries formed. The Eurasian plate motion shifted to the northwest along the Red River Fault zone (B) which helped shift regional plate motions.

The Philipine Islands (C) continued to be deformed by the northward movement of Australia. The Luzon arc collided with the Eurasian margin in Taiwan (D). Subduction occurred beneath Manila, Sangihe, and the Halmahera arcs (E) with new subduction beginning at the Negros and Philipine trenches (F). Strike slip faults connected these trench systems (G).

West Sundaland subduction and strike slip faulting (H) became segmented through Sumatra creating fore arc sliver plates. Extension along the fault system opened up the Andaman Sea basin (I) (**Figure 57**).

As subduction continued at the Eurasian-Philipine Sea-Australian triple junction (A), the Banda Sea (B) began to rift (**Figure 58**). The Mesozoic crust (C) north of Timor sank into the Java Trench. Australia migrated north towards the trench. Volcanic activity expanded the Banda arc eastward. Ceram (D) Volcanic activity expanded the Banda arc eastward. Ceram began moving east as subduction and strike slip faulting developed. The Molucca Sea continued closing from subduction occurring on both sides of the basin (E). The Sangihe fore arc began over-thrusting the Halmahera arc.

Lateral faulting along the Sorong Fault (between D & E) caused Tukang Besi to accrete onto Sulawesi locking the fault and creating a new splay south of the Sula platform (F). The Sula platform then collided with the east arm of Sulawesi causing a rotation to its present day position. Southward subduction of the Celebes Sea (G) began at the north Sulawesi trench.

The Woodlark basin (H) opened when collisions between the Caroline and New Guinea arcs shifted subduction of the Solomon Sea plate from south to north. Northward subduction pulled the ocean crust downward, ripping open the Papuan peninsular crust (**Figure 58**).

H. The Wharton Basin

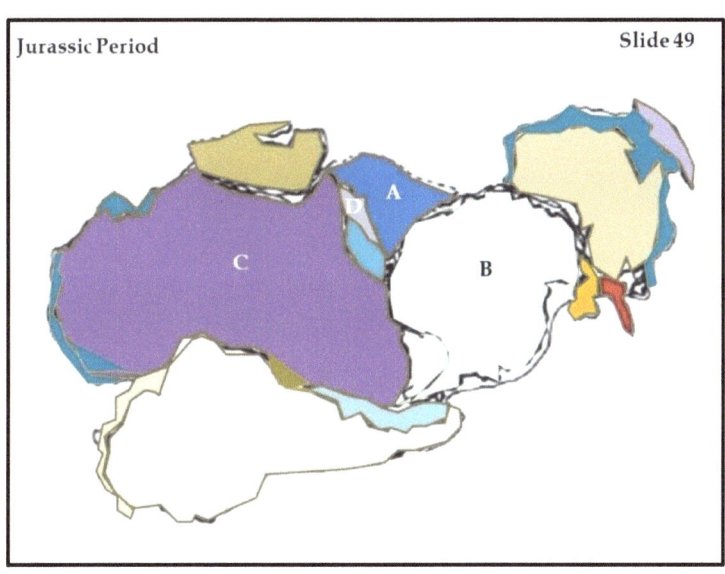

Figure 59. The Wharton Basin theory begins with the breakup of Pangea.

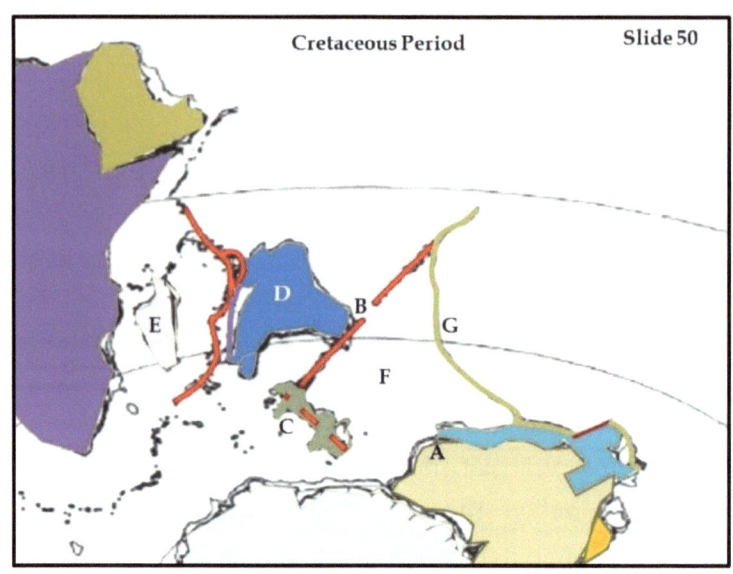

Figure 60. Ninety East Ridge.

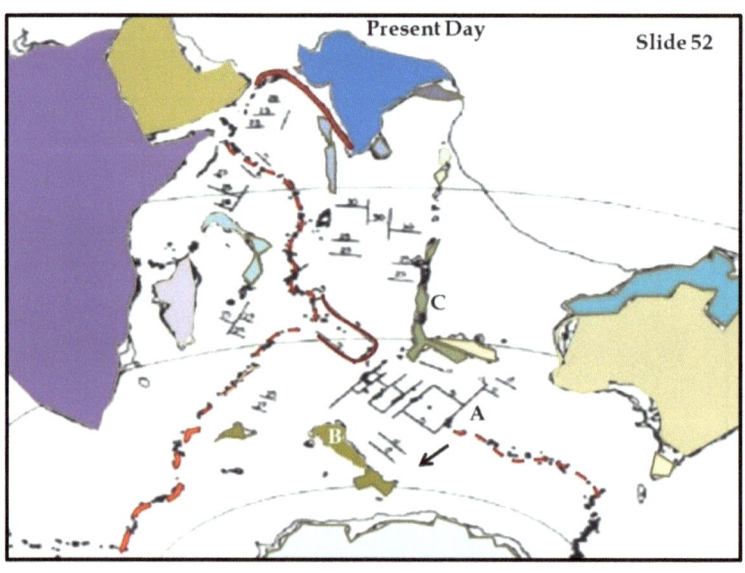

Figure 61. SE Indian Ridge &The Kerguelen Plateau.

Figure 62. Diamantina Fracture Zone.

The Wharton Basin tectonic history is complex and is not clearly understood. The basin history is based on stratigraphic data obtained from the Deep Sea Drilling Project. During the Jurassic Period, Gondwanaland was assembled with India (A) positioned in a north central wedge between Antarctica (B) and Africa (C). Madagascar (D) was wedged between Africa and India at this time (**Figure 59**).

During the Cretaceous Period 75 million years ago, Gondwana was proceeding to break up (**Figure 60**). Antarctica and Australia remained attached as a single continent (A). The Ninety East Ridge (B) formed as a transform fault joined with the Kerguelen Plateau (C).

107

India (D) and Madagascar (E) were already separated from Africa along a rift zone off the western Indian coast. Wharton Basin (F) was already formed between the Ninety East Ridge and the Australian-Antarctica continent. The Australian Plate as being pulled northward into the Sundaland Trench (G).

The formation of the SE Indian Ridge (A) separated the Kerguelen Plateau (B) from the Ninety East Ridge (C) (**Figure 61**). An existing hot spot remained beneath the Kerguelen Islands. The South Indian Ocean plate moved to the southeast.

Theories suggest the Diamantina Fracture Zone (A) was, and continued to be an old rift, embedded into the Indian Ocean plate (**Figure 62**). Between 130 and 45 million years ago, India separated from Antarctica and Australia. About 90 million years ago, Australia separated from Antarctica. Slow spreading occurred between 90 and 43 million years ago. The fracture zone formed as part of a rift zone which separated Australia from Antarctica and as Broken Ridge separated from the Kerguelen Plateau.

About 35 million years ago, the Ninety East Ridge (A) was beginning to rift apart from the Kerguelen Plateau (B). The SE Indian Ridge began to form (C). Broken Ridge (D) separated from the plateau. Australia separated from Antarctica drifting to the northeast (E) (**Figure 63**).

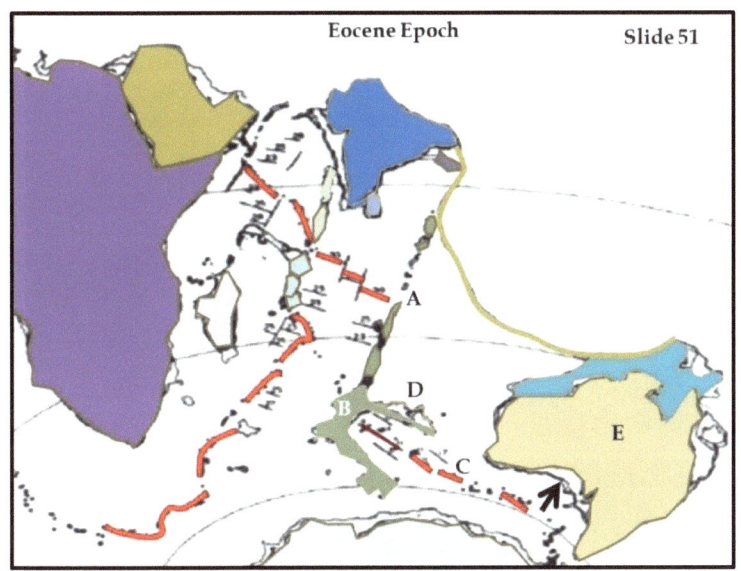

Figure 63. Ninety East Ridge Rift.

India separated from Madagascar about 83 million years, moving north as the Tethys Sea closed (**Figure 64**). Hot spots developed beneath the Bouvet, Marion, and Kerguelen Island Arcs forming the Algulhas Plateau located off the South African coast and the Madagascar Rise.

The Reunion hot spot (A) trace follows the Maldives (B) and Laccadives (C) Ridges, ending at the Deccan flood basalt province in western India. The Deccan basalts erupted between 68 and 66 million years ago. Shortly after the Deccan flood basalts erupted in western India (D), the Central Indian spreading ridge jumped from the Madagascar Basin (E) north to the western Indian margin (F). About 64 million years ago, the Seychelles Bank (G) belonged to the Deccan flood basalt province of western India.

I.   The Indian Ocean

Figure 64. India rifted from Madagascar.

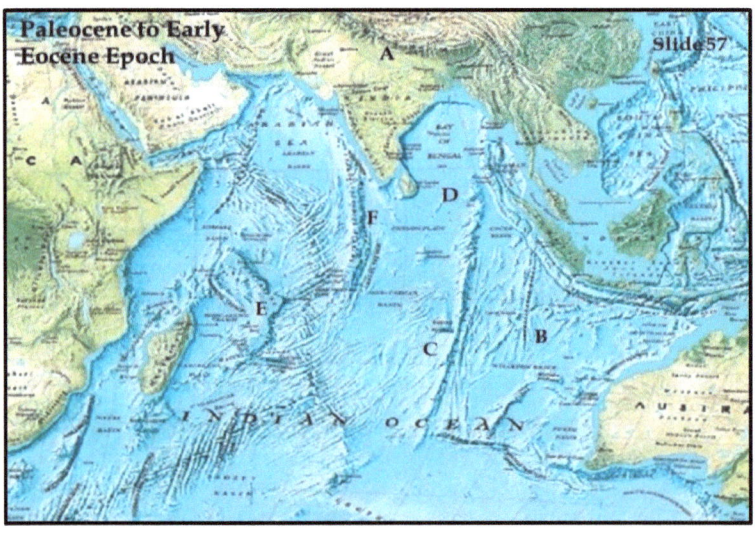

Figure 65. India-African-Eurasian collision.

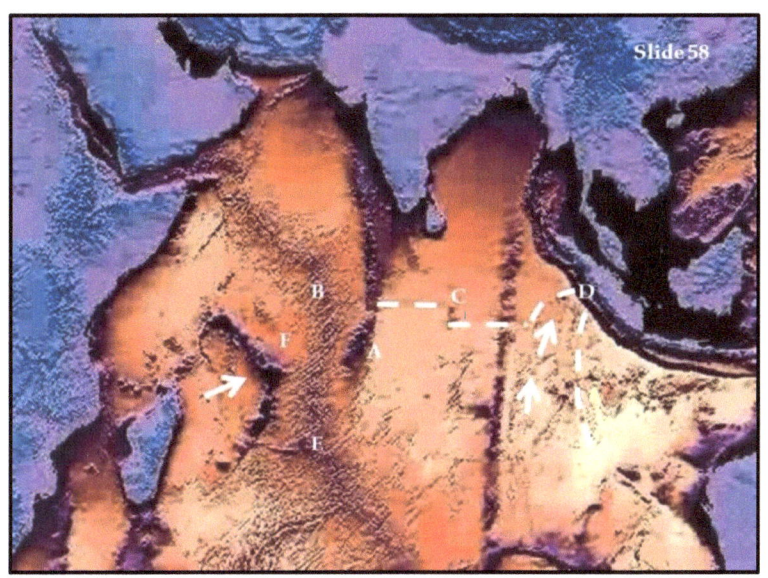

Figure 66. The Chagros Bank.

Figure 67. Central Indian Basin.

The Seychelles and a section of the Mascarene Plateau (H) rifted away from India accreting onto the African plate. About 56 million years ago, a small basin opened between the Seychelles and India. The Reunion hot spot began building a volcanic island (A). A long left lateral transform boundary was located west of the Reunion hot spot (I) about 56 million years ago. A transform junction connected the Arabian spreading center to spreading near and south of the Indian Ocean triple junction (**Figure 64**).

India and Africa collided with Eurasia, around 55 million years ago (**Figure 65**). Crustal compression took place along the collision boundary. The drifting of the Indian subcontinent slowed at this time (A).

The Australian plate (B) welded to the Indian plate, moving together as a single plate, 54 million years ago. The north central Ninety East Ridge formed when hot spots developed beneath the Bouvet, Marion, and Kerguelen islands.

The southern segment was built up by the Kerguelen hot spot around 47 million years ago (C). Spreading in the northeast Indian Ocean (D) ended. Spreading began to separate Australia from Antarctica. The Mascarene Plateau (E) and the Chagos-Maldives-Laccadives Ridge system (F) were formed as the northward continuation of the Reunion hot spot trace. Submarine eruptions built up the core of the ridge system into subaerial islands. The island chain was eroded back down to sea level, subsiding from oceanic crustal cooling.

More recent carbonate and reef deposits accumulated above the volcanic core (**Figure 65**).

The Chagos Bank (A) represents an older segment of the Reunion hot spot track which was separated from the younger Mascarene Plateau by post Eocene spreading of the Central Indian Ridge (B) (**Figure 66**). About 49 million years ago, the Reunion hot spot built up the northern Chagos Bank along a spreading ridge segment of the Central Indian Ridge. The Chagos Bank is presently extending from north to south.

Wiens and others proposed a model which separates the Arabian and Indian Plates from the Australian Plate by a diffuse plate boundary (C).

The plate boundary is oriented east to west, trending from the Central Indian Ridge near the Chagos Bank, to the Ninety East Ridge. From the Ninety East Ridge, the trend continues northward along the east side of the ridge to the Sumatra Trench (D). The ridge migrated northeast over the Kerguelen hot spot 38 million years ago, then over the Reunion hot spot 34 million years ago. About 34 million years ago, the Reunion hot spot was positioned beneath the Central Indian Ridge (E).

About 34 million years ago, the Kerguelen Plateau was built up during slow clockwise rotation of the Antarctica plate and the eastern Mascarene Plateau (F). The plateau was built up by northeast movement of the African plate.

The Ninety East Ridge became a left lateral fault along the eastern side of the ridge, moving the Wharton Basin northward. The fault represents the eastern boundary of the proposed diffused plate, thought to separate the Mid Indian Ocean plate from the Australian plate (**Figure 66**).

The Indian Ocean is thought to have formed between 10 and 5 million years ago (**Figure 67**). The Central Indian Basin (A) formed at the same time the Gulf of Aden (B) began to spread apart about 10 million years ago. Between 5 and 4 million years ago, the Aden spreading center expanded westward. The Central Indian Basin is currently being compressed from north to south (C). The western part of the Rodriguez Island Ridge formed between 10 and 8 million years ago.

Rodriguez Island marks the east end of a dike or sill lineation which extended from the main Reunion volcanic hot spot zone. The hot spot began erupting about 1.5 million years ago (D). Mauritius (E) formed by shield volcanic activity occurring between 8 and 7 million years ago, 3.5 to 2 million years ago, and 0.7 to 0.2 million years ago.

The younger volcanic phases may be related to underlying crustal doming and loading over the Reunion hot spot, followed by cooling and subsiding downstream from the melting zone.

Fracturing, resulting from bending and heating stresses within the crust, allowed smaller volumes of magma to rise from the upper mantle into the crust (**Figure 67**).

Chapter References

Chapter 1 Plate Tectonic References: Plate tectonics diagram published by the US Geological Survey. Posted on the USGS plate tectonics web site.

Earth Interior Model published by the USGS. Posted on the USGS plate tectonics web site.

Block models were redrawn from *National Geographic's Shaping of a Continent Map Series, 1985.*

Chapter 2 Paleogeographic References: Scotese, C.R., 2013. PALEOMAP PaleoAtlas for ArcGIS, Volumes 1-6, PALEOMAP Project, Evanston, IL (http:www.scotest.com). Pannotia map and later. Used with permission.

Chapter 3 References: The Caribbean Basin map sequences were redrawn from Pindell, J.L., and others, 1988. A plate kinematic framework for models of Caribbean evolution. Elsevier Science Publishers, Amsterdam, the Netherlands.

Figures 9 and 10 were obtained from the National Geographic Society Atlantic Basin (1965) and World Map (1981).

The East Pacific Basin map portion was cropped from the National Geographic Society Pacific Ocean Basin map, published in 1965.

Mediterranean reconstruction maps and text descriptions were interpreted from paleogeographic maps produced and published by Ron Blakely, Northern Arizona University.

Sequential paleogeographic drawings were recreated from Ron Blakely's set of paleogeographic reconstruction maps for the Mediterranean Basin, from the Early Jurassic Period through the Late Miocene Epoch.

Relief map of the Aegean Sea was published on the Wikipedia Commons web site. The map is used in accordance with posted terms and conditions.

Paleogeographic globe maps were obtained from Scotese, CR, 2013. PALEOMAP PaleoAtlas for ArcGIS, Volumes 1-6, PALEOMAP Project, Evanston IL. (http:www.scotese.com).

Pacific Ocean References: The World Ocean Map (1981), Atlantic Ocean Basin (1969), and Pacific Ocean Basin (1969) maps were published by the National Geographic Society.

Mediterranean Basin text was extracted from Kolker, A._. Geological history of the Central Mediterranean Basin: An outline of Cenozoic events. Published on the internet, no date provided.

Text for the Fiji Islands was obtained from the Fiji Ministry of Lands & Mineral Resources. Plate tectonic history of Fiji. ISSN 1016-2135, published on the internet.

Text for the Easter Islands was extracted from Neall, VE, Trewick, S.A. The age & origin of the Pacific Islands: a geological overview. Published on the internet by The Royal Society, 2008.

New Zealand text was extracted from The Geological History of New Zealand, University of Waikato, New Zealand. Published on the internet with no author citation.

The New Zealand satellite imagery was obtained from the WorldAtlas.com web site.

Philipine Plate text was extracted from Hall, R. 1998. The plate tectonics of Cenozoic SE Asia and the distribution of land and sea.SE Asia Research Group, Dept. of Geology, Royal Holloway University of London, Egham, Surrey. The set of plate reconstruction globes were extracted from the report appendices.

Wharton Basin drawings and text extracted from the Shipboard Scientific Party, Introduction and Explanatory Notes. Deep Sea Drilling Project. No authors cited.

Indian Ocean References: The 3D Map was obtained from an unidentified source posted on the internet.

Text was extracted from Wiens and others, 1985. A diffuse plate boundary model for Indian Ocean tectonics. Geophysical Research Letters, American Geophysical Union, Vol. 12, No. 7, pg. 429-432.

Duncan, RA. 1990. The volcanic record of the Reunion Hot Spot. Proceedings of the Ocean Drilling Program, Scientific Results, Vol. 115.

Ocean basin floor map was extracted from the National Geographic Society's World Map. 1981.

Other Titles Available in this book series

Introduction to Global Plate Tectonics I

Plate Tectonics Theory, Paleogeography,
& the Ocean Basins
By William A. Szary

Introduction to Global Plate Tectonics I: Plate Tectonics Theory, Paleogeography, and the Ocean Basins. Book I presents the basic theory for understanding the driving forces behind continental drift, and the opening and closing of the ocean basins. This is the first of a series of books which offers interested individuals and students a non-technical approach towards introducing *plate tectonics theory*, and the mechanisms for understanding the distribution of continents around the globe. *Paleogeography* is the study of reconstructing continental positions throughout geologic time, up to the present positions. Two maps are offered for predicting future continental positions as they may appear tens of millions years from now. The *Ocean Basins* presents the geologic history of the major oceans and how they formed as continents drifted apart, collided, and reassembled throughout geologic time. Video slide shows are available for viewing on Slideshare.com (free of charge). Enter search word Category Education, keyword plate tectonics. The slide show offers basic animation, audio text narrative, and music soundtrack. List Price: $24.18. Available from Amazon.com.

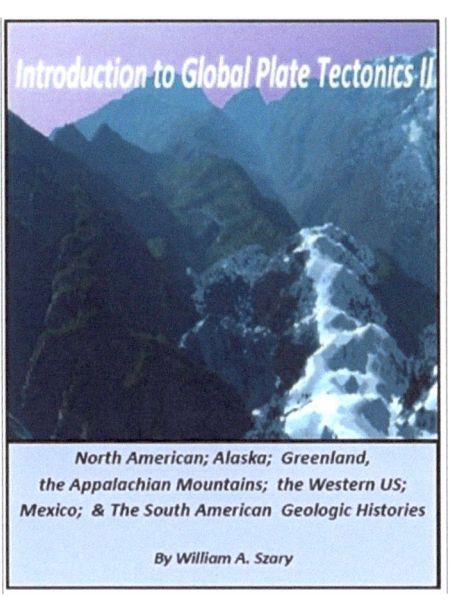

North American; Alaska; Greenland,
the Appalachian Mountains; the Western US;
Mexico; & The South American Geologic Histories

By William A. Szary

Introduction To Global Plate Tectonics II: North American Tectonic History. Book II presents the Precambrian Craton & Phanerozic Geologic History of the North American craton from Early Precambrian time, greater than 2 billion years ago, to the Early Paleozoic Era Cambrian Period, 530 million years ago. The tectonic assembly of Alaska, the Appalachian Mountains, the Western US, Greenland, Mexico, and South America are discussed. The book utilizes many picture maps, accompanied by text discussion, to present the subject matter in an understandable presentation. Video slide shows are available for viewing on Slideshare.com (free of charge), Category Education, keyword plate tectonics. The slide show offers basic animation, audio text narrative, and music soundtrack for the North American Precambrian and Phanerozoic Eras. South American, Alaska, Mexico, and Greenland are not included with the presentation. Book List Price: $41.68. Available from Amazon.com.

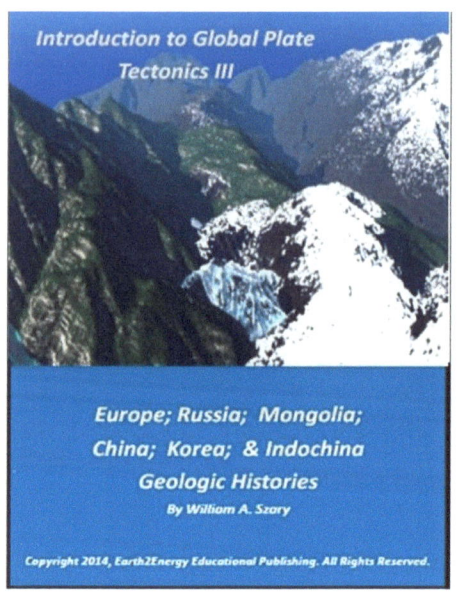

Introduction To Global Plate Tectonics III: Europe, Russia, Mongolia, China, Korea, and Indochina. _Introduction To Global Plate Tectonics III_ is the third part of a five part series covering the subject of plate tectonics, paleogeography, and the drifting and build out of continents. Each book was designed as a picture guide containing many images extracted from scientific journal articles written by research professors on the subject matter. The text content has been rewritten to help explain technical terms, converting the terms into a more understandable language for those interested in learning about geological sciences, but have not yet mastered the terminology. Images were redrawn by adding color and texture to highlight key events explained by the text.

Part III covers the development of the European basement continuing through to the present day. Chapters on Russia, Mongolia, China, Korea, and Indochina geologic histories are included Book III. This book focuses on the more technical aspects of continental development. Book List Price: $34.33.

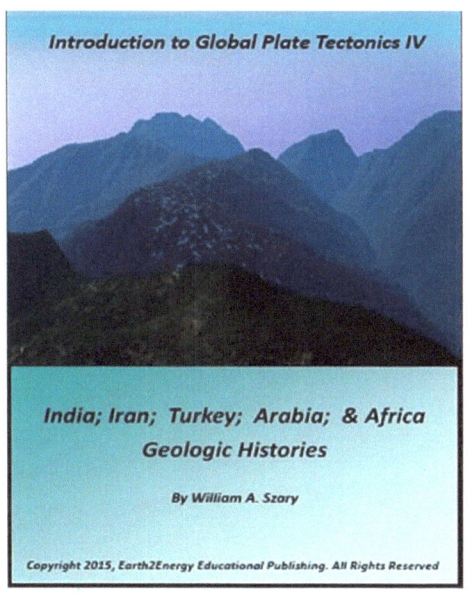

Introduction To Global Plate Tectonics IV is the fourth part of a five part series covering the subject of plate tectonics, paleogeography, and the drifting and build out of continents. Each book was designed as a picture guide containing many images extracted from scientific journal articles written by research professors on the subject matter. The text content has been rewritten to help explain technical terms, converting the terms into a more understandable language for those interested in learning about geological sciences, but have not yet mastered the terminology. Images were redrawn by adding color and texture to highlight key events explained by the text.

Part IV covers the development of the Indian, Iranian, Turkish, Arabian, and African basement continuing through to the present day. This book focuses on the more technical aspects of continental development. Book List Price: $40.63. Available from Amazon.com.

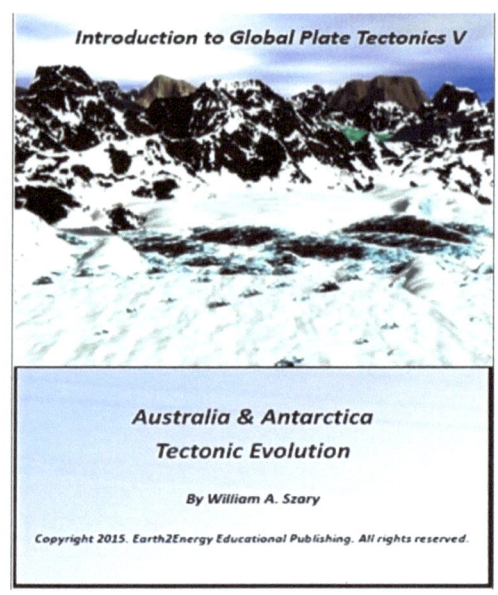

Introduction To Global Plate Tectonics V is the fifth part of a five part series covering the subject of plate tectonics, paleogeography, and the drifting and build out of continents. Each book was designed as a picture guide containing many images extracted from scientific journal articles written by research professors on the subject matter. The text content has been rewritten to help explain technical terms, converting the terms into a more understandable language for those interested in learning about geological sciences, but have not yet mastered the terminology. Images were redrawn by adding color and texture to highlight key events explained by the text.

Part V covers the development of the Australia and Antarctica Precambrian Era basement continuing through to the present day. This book focuses on the more technical aspects of continental development. Book List Price $16.48. Available through Amazon.com.

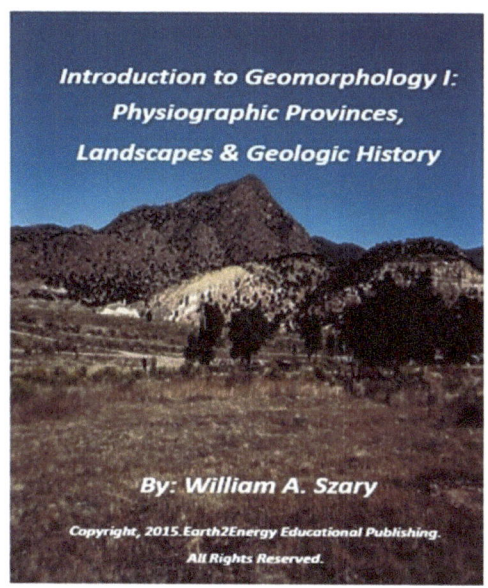

Landscapes and landforms are expressions of geologic structure, processes, and time. Each landscape expresses certain physical similarities grouped into provinces based on geologic history, geologic structure, rock formations, and the processes which weather and erode landscapes based on climatic conditions. All factors play a role in shaping and sculpting landscapes and landforms.

Introduction to Geomorphology I is a review of the geologic history behind each physiographic province, providing typical and atypical photographic representations for each province recognized in continental United States (125 photographs).

Introduction to Geomorphology II presents structures, landforms, and geologic processes presenting constructional (tectonic and volcanic landforms), destruction (weathering processes), erosional processes (mass wasting, hill slope evolution), fluvial (river), and coastal processes using many photographic examples for each subject presented. Book List Price: $34.68. For sale on Amazon.com.

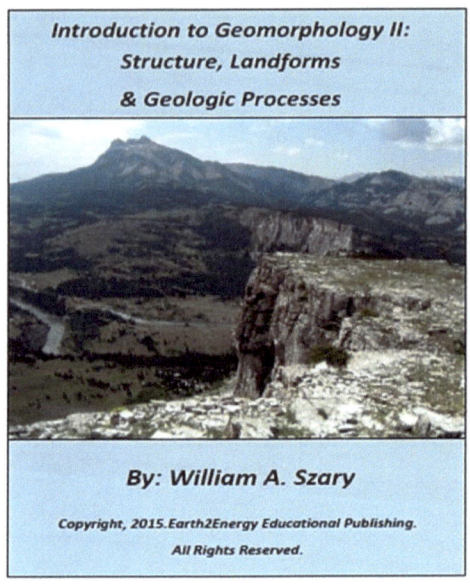

Book II continues with expanding the description of geomorphic provinces describing landscapes in the context of geologic structure, landforms, and the basic principles of processes which shape landforms and landscapes. Book I addressed the geomorphic provinces in terms of the geomorphic province, landscapes and geologic history behind each province.

Many photographs are presented in this book covering constructive, destructive, mass wasting, fluvial (river) processes, and coastal processes. Photographs were obtained from public domain sources. Author and source are provided with the appropriate photograph. Where no citations are provided, photographs were taken by the author.

Both books in the Geomorphology series are intended for those interested in earth sciences at the secondary school, community college, and first year undergraduate level of study. Price: $34.68. Available from Amazon.com.

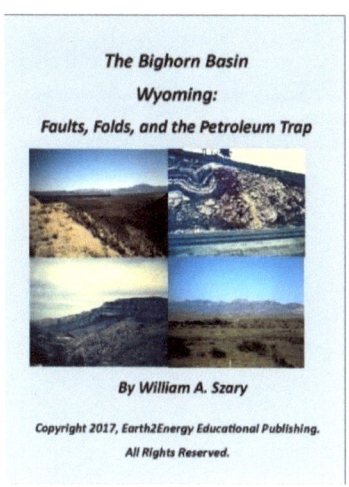

The Bighorn Basin, Wyoming: Faults, folds, and the Petroleum Trap. Wyoming consists of a set of basins separated by mountain elts uplifted during the Laramide orogeny. This book focuses on the Bighorn Basin, it's geologic structure and petroleum resources. The basin is located in the north central part of the state. Faults and folds are covered over by younger sedimentary rock sequences which allowed the basin to become a major petroleum producer in the western US. This book sets out to address the Bighorn Basin petroleum province through discussion on regional and basin specific geologic history, structure, and stratigraphy.

Chapter 1 summarizes Western regional US geologic history. Chapter 2 presents Bighorn Basin stratigraphy, sedimentation, and depositional environments. Chapter 3 summarizes the basin's geologic setting. Chapter 4 presents basic topics covering petroleum geology and significant structures associated with the Bighorn Basin. Petroleum provinces are presented in Chapter 5.

Book price: 29.99. Available through Amazon.com.

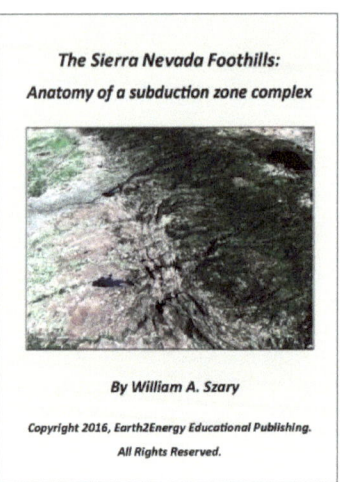

The Sierra Foothills were formed during the Mesozoic collision with the Paleozoic Shoo Fly Complex when a series of small island arcs and spreading ridge centers accreted and subducted beneath the continental margin of the Sierra Nevada during the Nevadan orogeny. This book presents the regional geologic setting of the Sierra Nevada Foothills terrance, offering tectonic models for oceanic plate-continental collisions followed up by geologic evidence collected from the New Hogan Reservoir for support. Tectonic models are proposed to help understand the subduction zone processes and terranes where chaotic conditions prevail. Book price: $59.99. Available from Amazon.com.

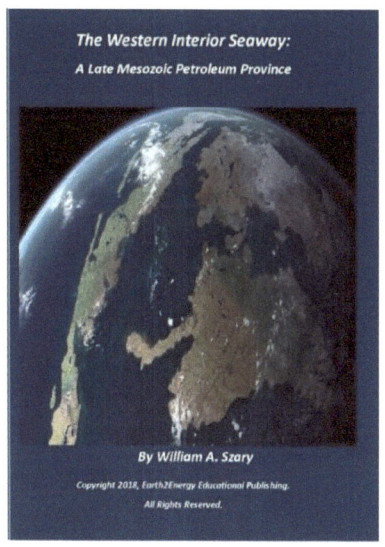

The Western Interior Seaway:
A Late Mesozoic Petroleum Province

By William A. Szary

The Western Interior Seaway accumulated thick sediments which support hydrocarbon sources and reservoirs beginning in the Jurassic Period where the seaway was limited to the Canadian British Columbia and Saskatchewan Provinces. Seas retreated and advanced throughout the North Canadian Provinces up until the Early Cretaceous Period when the seaway began to invade the western US states. By Late Cretaceous time, the seaway was at its maximum coverage throughout the Great Plains and along the Rocky Mountain front. The western parts were subjected to deformation while the eastern part remained undisturbed. Deformation along the Rocky Mountain front promoted fluvial floodplain and swamp deposition along the western shoreline building out deltaic deposits as the drainage reached the western shoreline. Coal deposits accumulated. The eastern shoreline consisted mostly of terrestrial deposits reworked as they invaded the seaway. Marine limestone accumulated in the quieter waters while sandstone, shale, and mudstone buried limestone under similar conditions. These sediments promoted hydrocarbon development.

Chapter 1 provides and overview of the4 western US Mesozoic and Cenozoic geologic history. Chapter 2 describes the stratigraphy and sedimentation associated with the seaway during transgressive-regressive cyclic deposition. Chapter 3 presents coal and petroleum resources of Canada tied to the Western Canadian Sedimentary Basin. Chapter 4 presents petroleum provinces in the western US developed by the Western Interior Seaway. Chapter 5 summarizes the paleogeography and hydrocarbon development. Book price: $49.99. Available from Amazon.com.

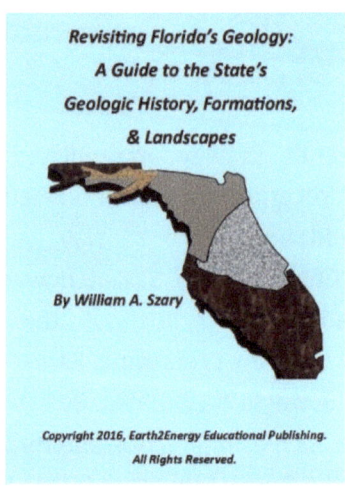

Revisiting Florida's Geology:
A Guide to the State's
Geologic History, Formations,
& Landscapes

By William A. Szary

Copyright 2016, Earth2Energy Educational Publishing.
All Rights Reserved.

Revisitng Florida's Geology presents and assembly of the state's early geologic history during the Paleozoic into the Mesozoic Era including a discussion on geologic formations, structures, and tectonic processes forming the early basement complex. The Cenozoic history is presented in context of the uppermost limestone and sedimentary rock formations.

The various types of landscapes are presented using selected county geologic maps to show which formations are responsible for producing flatlands, gently to moderately sloping hills, steep hills, valleys, and other features typically observ3ed while traveling throughout the state. The book presents 294 drawings and photographs to guide the reader through the text.

Chapter 1 provides a brief review of Florida's Paleozoic through Mesozoic Era basement geology. Chapter 2 provides a brief summary of Florida's geologic structures and history. Chapter 3 provides a summary of Florida's Cenozoic Era geologic formations. Chapter 4 provides an overview of five of Florida's most common native minerals. Chapter 5 presents Florida's surficial geology and associated landscapes along with a discussion of Florida's marine and fluvial marine shoreline terraces marking the various sea level stands throughout the state and the relationship to surficial geology for selected counties. Chapter 6 presents a photographic presentation on Florida's karstic processes and landforms. Book Price: $59.99. Available on Amazon.com.

For additional information contact William Szary at wszary@netzero.net.

Earth2Enegy Educational Publishing
Port Richey FL 34668
Earth2Energy is a registered trademark.

www.ingramcontent.com/pod-product-compliance
Lightning Source LLC
Chambersburg PA
CBHW040807200526
45159CB00022B/39